The Structures and Properties of Solids
a series of student texts

General Editor:
Professor Bryan R. Coles

The Structures and Properties of Solids 4

The Electronic Structures of Solids

B. R. Coles, B.Sc., D.Phil.
Professor of Solid State Physics, Imperial College
of Science and Technology, London

A. D. Caplin, M.A., M.Sc., Ph.D.
Senior Lecturer in Physics, Imperial College
of Science and Technology, London

Edward Arnold

© B. R. Coles and A. D. Caplin 1976

First published 1976 by Edward Arnold (Publishers) Limited
25 Hill Street, London W1X 8LL

Boards Edition   ISBN: 0 7131 2526 8
Paper Edition   ISBN: 0 7131 2527 6

Typeset in Great Britain by
Preface Limited, Salisbury
and Printed by
J. W. Arrowsmith Ltd., Bristol

Preface

Our object in this book has been to supply the student of solid state physics
with the essential concepts he will need in considering those properties of solids
that depend primarily on their electronic structures, and some idea of the
electronic character of particular materials and groups of materials. There is a
general tendency for accounts of the electronic structures of solids to begin with
the free-electron gas and then to go on to consider its modification by the
periodic lattice. We believe this under-emphasizes the relationship between
electron states in solids and those in atoms and molecules. In an effort, therefore,
to redress the balance we have devoted the first chapter to electron states in
atoms, concentrating especially on many-electron atoms of the sort that make
up solids of interest rather than on a detailed discussion of the hydrogen atom,
and then in the second chapter we have discussed bonding of atoms in molecules
that leads naturally to discussions of solids. Chapter 3 deals with the free-electron
gas, indicating both the applications and the limitations of this model, and in
Chapter 4 we deal at some length with the descriptions of electron states in
crystals. We have tried to limit the formal discussion of electron states in
periodic lattices to those aspects which we need in discussing the general
properties of solids. The later sections of Chapter 4 then examine the features
of the electronic structures of particular solids on which their special characteristics
depend.

   The next volume in this series (by Professor J. S. Dugdale) will deal in detail
with electron transport in solids, but we felt it appropriate in the present book to
indicate the different ways in which electrons in solids respond both to steady
electric fields (in their transport properties) and to electromagnetic radiation (in
their optical properties). In Chapter 5 these matters are discussed, especially in
relation to the essential distinctions between metals, semiconductors and
insulators. In the last chapter we look at some aspects of solids that cannot be
adequately discussed within the electron gas and simple periodic lattice models.
The list of such special topics is by no means complete, but we have tried to
introduce the reader to the main concepts involved in modern accounts of
disordered materials, metal—insulator transitions and superconductors. We have
made no attempt to deal with those aspects of solids which depend on spin-
dependent interactions between electrons or between electrons and partly

filled ion core states; such aspects lie more properly within the scope of the forthcoming volume on *The Magnetic Properties of Solids* by Dr. J. Crangle.

The level is essentially that appropriate to students in the second or third year of an undergraduate physics course, but undergraduate electrical engineers, metallurgists and material scientists who are prepared to take some results of quantum mechanics on trust should also be able to follow the general lines of the discussion.

London   BRC
1975   ADC

Contents

5   METALS, INSULATORS AND SEMICONDUCTORS

6   SPECIAL TOPICS

List of Symbols

a	lattice spacing
a_0	Bohr radius
$\mathbf{a, b, c}$	primitive translational vector
d_{hkl}	crystal plane spacing
D	heat of dissociation
e	electronic charge
E	total energy
E_g	energy gap (between bands)
$f(E)$	Fermi–Dirac distribution function
G	reciprocal lattice vector
h	Planck's constant
$\hbar$	$h/2\pi$
$\mathcal{H}$	Hamiltonian operator
ΔH_s	Heat of sublimation
h, k, l	Miller indices
$\mathbf{i, j, k}$	Cartesian unit vectors
J (chap. 1)	total angular momentum quantum number for an atom
J (chap. 3)	current density
k	wave-vector
k_F	Fermi wave-vector
k_B	Boltzmann's constant
l	orbital angular momentum quantum number
L	total orbital angular momentum quantum number for an atom
L (chap. 3)	crystal size (side of cube)
m	electronic mass
m^*	effective electronic mass
m_l	z-component angular momentum quantum number
n	first quantum number
$n(E)$	density of states
N	total number of electrons
p	momentum
r_w	radius of Wigner–Seitz sphere
S	total spin angular momentum quantum number for an atom

dS	surface element
T	lattice translational vector
T	absolute temperature
T_c	superconducting transition temperature
v_g	group velocity
V	potential
Z	atomic number
α, β, γ	primitive reciprocal lattice vectors
γ	electronic specific heat coefficient
Δ	energy gap (across zone planes)
ϵ	one-electron energy
$\mathcal{E}$	electric field
κ	dielectric constant
λ	spin-orbit coupling parameter
μ	chemical potential
ν_1, ν_2, ν_3	integers
ρ	electrical resistivity
σ	electrical conductivity
τ	relaxation time
$d\tau$	volume element
ϕ	atomic wave function
ψ	wavefunction (one electron)
Ψ	wavefunction (many electron)
χ	magnetic susceptibility
ω	phonon or optical frequency
Ω_{BZ}	Brillouin zone volume

The Periodic Table of the Elements

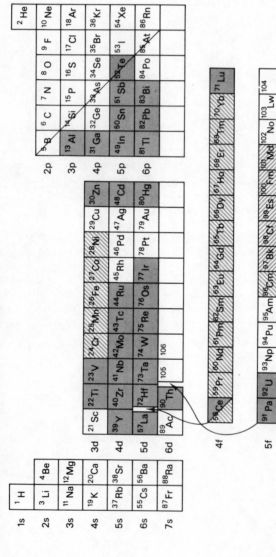

The symbols 2p, 3d, etc. indicate that that sub-shell begins to be occupied in atoms of the element to the right of the symbol; the later occupation of the levels is not always uniform however.

The diagonal line indicates approximately the division between metallic and non-metallic behaviour.

Elements indicated thus ▨ possess a magnetically ordered solid phase at some temperature.

Elements indicated thus ■ possess a superconducting solid phase at some temperature and pressure fairly easily available with modern techniques.

Elements indicated thus ▧ are too rare or unstable for their character to have been established but they are expected to yield magnetic order.

1

Electronic Structure of Atoms

1.1 Introduction

In all discussions of the properties, including the structures and binding energies, of molecules and solids we need as a starting point a description of the electronic structure. In principle this description is contained in a wavefunction which is a function of the coordinates of all the electrons involved in the bonding process. In practice we shall adopt the procedure of the atomic physicist when he deals with the many-electron atom, and use one-electron wavefunctions; that is to say we shall think in terms of a wavefunction describing an electron that moves in the potential due to all the nuclei and all the other electrons in the system. We shall deal in Section 1.3 with the detailed way in which the atomic physicist tries to make sure that the wavefunctions he uses for electrons in many-electron atoms are self-consistent, that is, capable of yielding (with the nucleus) the potential he assumes them to be subjected to. In Chapters 3 and 4 we shall be using wavefunctions of this one-electron sort for electrons in solids, although there we shall simplify the situation by lumping together as an ion core potential the potential due to the nuclei and that due to all the electrons in closed inner electron shells that are not modified by bringing the constituent atoms together in the solid. Such an ion core potential is useful, of course, for some atomic situations also, as when one brings out the similarities of the atomic spectra of the alkali metals (Na, K, etc.) and their relationship to that of the hydrogen atom by thinking of the outermost electron as moving in the spherically symmetric potential of Na^+, K^+, etc.

At normal interatomic distances in solids there is considerable overlap between the atomic one-electron wavefunctions of the outer electrons, but the modification to these wavefunctions close to the individual nuclei will be very small and it is often useful to label these electron states with the notation appropriate in the isolated atoms. Since, as we have seen, the atomic physicist describes those atomic states in terms of a spherically symmetric (central) potential he is able to classify them using the notation that spectroscopists have used, and quantum mechanics has justified, for the hydrogen atom.

It is therefore appropriate to begin our description of electronic structures by recapitulating that notation and relating it to the spatial distribution of the charge clouds of the corresponding one-electron wavefunctions. The shapes of these wavefunctions play a vital role in governing the structures of molecules and of many elemental solids, and even in metals where the shapes are not normally considered the degree of overlap from atom to atom of different groups of electrons is central to the different roles they play in the properties.

There are properties, of both atoms and solids, in which the collective behaviour of groups of electrons is important, and our one-electron wavefunctions will then no longer be an adequate starting point for the discussion of those properties. In Section 1.4 we will introduce some of the appropriate descriptions when the angular momentum (rather than the one-electron energy) of groups of electrons is important. Such descriptions will turn out to be vital in accounting for the magnetic properties of the more magnetically interesting solids.

1.2   The hydrogen atom

The simple structure of the H atom, a single electron moving in the Coulomb potential of a single proton, makes possible a more rigorous application of quantum mechanics than for any other situation in physics, and we shall assume that the reader has gone through some of the mathematical exercises involved.

The solutions to the Schrödinger equation (the eigenstates of the Hamiltonian) that describe allowed energy levels are labelled with three integer quantum numbers n, l, and m_l that specify the radial extent of the wavefunction, the orbital angular momentum and the component of that angular momentum in some reference direction. We shall refer to a state specified in this way as an *atomic orbital*, and the purely Coulomb potential of the H atom ($V = -e^2/r$) introduces the simplifying feature, lost in all other atoms, that to a good approximation all the orbitals of a given n have the same energy (are degenerate) given by $E_n = -(1/n^2)Z^2(me^4/2\hbar^2)$ with $Z = 1$.

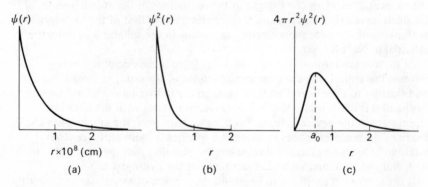

Figure 1.1   The ground state (1s) wavefunction of the hydrogen atom.
(a) $\psi(r)$; (b) $\psi^2(r)$; (c) $4\pi r^2 \psi^2(r)$.

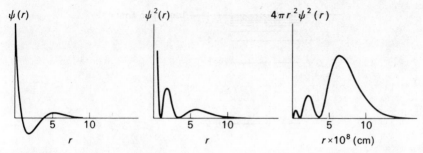

Figure 1.2 The 3s wavefunction.

In spherical polar coordinates the wavefunctions can be written in the form

$$\psi(r, \theta, \phi) = R(r)\Theta(\theta)\Phi(\phi)$$

and for any central potential the angular parts of the solutions can be written in the compact form

$$\Theta(\theta) = P_l^{m_l}(\cos \theta) \quad \text{where} \quad 0 \leqslant l \leqslant n$$

$$\Phi(\phi) = e^{im_l\phi} \quad \text{where} \quad -l \leqslant m_l \leqslant l$$

where $P_l^{m_l}(\cos \theta)$ is an associated Legendre polynomial.

The lowest energy orbital has no angular dependence of ψ ($l = m_l = 0$) and is simply the $n = 1$ state with wavefunction

$$\psi(r) = [1/\sqrt{(\pi a_0^3)}] \exp(-r/a_0)$$

where a_0 is the Bohr radius $\hbar^2/me^2$ which is numerically equal to $0\cdot529 \times 10^{-10}$ m and is a useful atomic unit of length. The factor $1/\sqrt{(\pi a_0^3)}$ is the normalizing factor required to make the integrated probability density $\int \psi^*\psi d\tau$ equal to unity if taken over all space. Usually we shall not include such normalizing factors in our expressions for wavefunctions. Since this ground state wavefunction is wholly real we can plot it as a function of r. Also, because $\psi^*(r)\psi(r)d\tau$ (in this case $\psi^2(r)d\tau$) is the probability of finding the electron in a small volume $d\tau$ at the point r) we can plot ψ^2 and also the quantity $\psi^2 4\pi r^2$, which has the property that $\psi^2 4\pi r^2 dr$ is the probability that the electron is at a distance r (i.e. in a spherical shell of thickness dr and surface area $4\pi r^2$) from the nucleus. These quantities are shown in Fig. 1.1.

All eigenstates with $l = 0$ have a cusp in ψ at the nucleus, and in all the radial dependence contains a factor $\exp(-r/na_0)$. They also have $(n - 1)$ spherical nodes, that is, points where ψ crosses the r-axis. Fig. 1.2 shows the character of $n = 3$ orbital of $l = 0$. Details of hydrogenic wavefunctions are given in Appendix A.1.1 to this chapter, and the resultant energy level scheme is shown in Fig. 1.3. Because of the exponential fall-off no well-defined boundary of the charge cloud

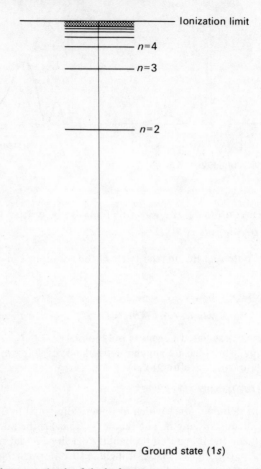

Figure 1.3   The energy levels of the hydrogen atom.

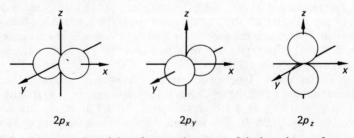

Figure 1.4   Representation of three 2p states by means of the boundary surface.

corresponding to an orbital exists, but it is often useful to specify a surface (a sphere for $l = 0$) of constant $| \psi |$ such that only 10% of $\int \psi^* \psi d\tau$ lies outside it. We call this the *boundary surface*.

From this point on we shall adopt the usual convention of referring to $l = 0, 1, 2, 3, 4 \dots$ orbitals as s, p, d, f, g $\dots$ orbitals, so that Fig. 1.2 shows the 3s-orbital.

For states of $l > 0$ we have to consider also the angular dependence. The wavefunctions for $n = 2$ and $l = 1$, the 2p functions, are

$$\psi_{2p}^0 = R(r)\cos\theta$$
$$\psi_{2p}^{\pm 1} = R(r)\sin\theta \, \exp(\pm i\phi)$$

An alternative basis set whose members are still eigenstates of the Hamiltonian if an external electric field of high symmetry is present (although not all eigenstates of the operator that gives the values of the z-component of the orbital angular momentum) can be chosen as

$$\psi_{2p}^x = (x/r)R(r) = R(r)\sin\theta\,\cos\phi$$
$$\psi_{2p}^y = (y/r)R(r) = R(r)\sin\theta\,\sin\phi$$
$$\psi_{2p}^z = (z/r)R(r) = R(r)\cos\theta$$

since

$$(x \pm iy)R(r)/r = R(r)\sin\theta\,\exp(\pm i\phi).$$

These three functions (the $2p_x$, $2p_y$ and $2p_z$) functions can now be sketched as a function of x, y and z and we can make three-dimensional representations of the boundary surface (as defined above) which give an idea of the shape of the charge cloud and are particularly appropriate for p-states in cubic crystals. These are shown in Fig. 1.4. (Notice that the superimposed charge clouds of the three possible p-orbitals give an electron density proportional to $(x^2 + y^2 + z^2)R^2(r)$ and therefore spherically symmetric. This is the case for any complete subshell of orbitals of given n and l.)

A similar alternative set of orbitals capable of having their wavefunctions simply written in cartesian coordinates can be found for the 3d (i.e. $n = 3, l = 2$)-orbitals; the angular parts of the wavefunctions now yield factors proportional to $xy, yz, zx, x^2 - y^2$, and z^2 respectively.

In the above discussion we have not mentioned the final quantum number, the spin; but the Pauli principle allows values of m_s, the quantum number specifying the z-component of the electron spin angular momentum, to take values of either $+\frac{1}{2}$ or $-\frac{1}{2}$, and the total number of allowed states of an electron in a H atom is twice the number of orbitals specified by n, l, and m_l. Since the number of orbitals for a given l is equal to $(2l + 1)$ and l may take values $0, 1, 2 \dots (n - 1)$ the number of allowed states for an electron in a particular shell of orbitals (specified by a given n) will increase in the order 2, 8, 18, 32, $\dots$

and we shall see later the significance of these shells and subshells (all the orbitals of given n and l) for the periodic table of the elements.

1.3   The many-electron atom

Probably many readers will be familiar with the idea that the arrangement of the periodic table is explicable in terms of the sequential filling of energy levels. Suppose, for example, we start with a bare Na nucleus ($Z = 11$); it will have a set of energy levels for electron occupation that is identical with that for the H atom (see Fig. 1.3), except that the energy scale is Z^2 times greater. If we now add the 11 electrons, and ignore the interaction between them, then it is natural that they should occupy the 11 lowest energy states of this hydrogenic system: two electrons, one of each spin, go into the 1s-level, two into the 2s-level, six into the 2p-levels, and the final electron occupies one of the 3s-states. This listing of the occupancy of successive shells is known as the *configuration*, and is written as $(1s^2)(2s^2)(2p^6)3s$; the parentheses indicate a full, or *closed*, subshell.

We shall see later that the detailed structure of the periodic table is explained by the fact that the energy level scheme has a slightly different order for most atoms from that of Fig. 1.3. In the H atom the energy is determined solely by the principal quantum number n; in many-electron atoms states of the same n do have similar energies, but the energy now depends upon the value of l too.

This simple picture of many-electron atoms does, with the modification just mentioned, put the energy levels in the correct *order*, but completely fails to predict the energies themselves. If we were to take the model seriously, the first ionization energy for Na would be 11^2 times the energy required to ionize a 3s-electron from H, that is $11^2 \times 13 \cdot 6/3^2 = 183$ eV, whereas the experimental value is only $5 \cdot 14$ eV. The reason is that we have ignored the electron—electron repulsion, which to a large extent cancels the electron—nuclear attraction; in a Na atom each electron is attracted by a positive charge of 11, but is repelled by 10 other negatively charged electrons.

Dynamical problems involving more than two interacting particles are impossible to solve exactly (in either classical or in quantum mechanics), but certain features of the solution can be found immediately in quantum mechanics and it is these that allow considerable simplification of the problem. Suppose we fill up the 11 lowest-lying states of a Na atom with non-interacting electrons, and then slowly re-introduce the Coulomb repulsion between electrons. For the moment focus attention on the electron occupying the 3s-orbital; it sees the nuclear charge plus the (negative) charge of all the other electrons; however those electrons form closed shells, and the charge distribution of a closed shell is spherically symmetric. There is no assymmetry in the electron—electron interaction that can destroy the sphericity of the charge associated with those closed shells, and so the potential in which the 3s-electron finds itself is still necessarily spherically symmetric, and we still have a central field; in that case orbital angular momentum must be conserved, and correspondingly the orbital angular momentum quantum numbers l and m_l, and incidentally the spin quantum number m_s, are still good ones. The orbital quantum numbers control the angular dependence of the wavefunction ($\S 1.2$), so that too can be found immediately; in this case, with an s orbital, the wavefunction itself is spherically symmetric, and likewise the charge density.

The radial dependence of the wavefunction is a bit harder to follow as the electron—electron repulsion is turned on. Initially, in the absence of interaction, the characteristic radius of a state of radial quantum number n in the field of a nucleus of charge $+Ze$ is of order na_0/Z so that (Fig. A.1.1.1, p. 18) a 3s-electron spends most of its time outside the 1s-, 2s- and 2p-electrons. Consequently, for distances large compared with the radial extent of the closed-shell electrons, the potential seen by the 3s-electron is simply that of a *single* positive charge, for the other electrons have *screened* it from the nucleus. Because the effective attractive charge Z_{eff} is that much less, the radius of the 3s-state increases, and the screening becomes that much better. To a first approximation the 2s- and 2p-electrons are screened from the nucleus by only the two 1s-electrons, so that they see an effective nuclear charge of +9, and the 1s-electrons themselves are unscreened and feel the full nuclear charge. We can then calculate the energies of these electrons (in eV) from

$$E_{scr} = -Z_{eff}^2(13 \cdot 6/n^2)$$

and the answers come considerably closer to experiment than did the unscreened estimates. For example, for the 3s-valence electron of Na, the ionization energy, with $Z_{eff} = 1$, is now $1 \cdot 5$ eV. To do better than this, it is necessary to calculate the actual charge distribution of the closed-shell electrons, and obtain from it a detailed electrostatic potential (which is no longer of $1/r$ form) for the 3s-electron, and so calculate its wavefunction. It is also necessary to allow for the effect of the 3s-electron on the closed-shell electrons, and evidently there is a self-consistency problem. This approach, which has been very successful in atomic physics, is known as the Hartree self-consistent field treatment, and the procedure for calculating the one-electron energy levels of an atom of atomic number Z, is as follows:

(i)   Guess a set of wavefunctions for $(Z - 1)$ electrons being guided by the probable configuration of the atom.

(ii) Construct the potential due to the nucleus $(-Ze^2/r)$ and $(Z - 1)$ electrons in states described by these wavefunctions.

(iii) Solve the Schrödinger equation for the Zth electron moving in this potential, obtaining enough solutions to accommodate $Z - 1$ electrons at the lowest possible energy in accordance with the Pauli principle. (This filling up in order of the one-electron levels is called the Aufbau principle.)

(iv)   Use these one-electron states as in (ii) to reconstruct the potential.

(v)   Repeat step (iii);

and so on until self-consistency is achieved, that is, until the one-electron levels found for a particular potential are just those which will, in conjunction with the nuclear charge, give rise to that potential.

When self-consistency is achieved the expressions for the potential V_i, the appropriate Schrödinger equation, and the Hamiltonian $\mathcal{H}_i$ for the ith electron are as follows, where ϵ_i is called the one-electron energy:

$$V_i = -\frac{Ze^2}{r} + e^2 \sum_{j=1 \neq i}^{Z} \int \frac{|\psi_j(r_j)|^2}{r_{ij}} \, d\tau_j \quad \text{where } r_{ij} = |\, r_i - r_j \,|$$

$$\mathcal{H}_i \psi_i(r) = \epsilon_i \psi_i(r)$$

$$\mathcal{H}_i = -\frac{\hbar^2}{2m} \nabla_i^2 + V_i(r)$$

This procedure is of course extremely laborious and the numerical calculations involved were among the earliest important achievements in physics of numerical computers.

An approximate expression for the total electronic energy of the atom will be given by

$$E \simeq \sum_i \epsilon_i - \frac{1}{2} \sum_{i \neq j} \iint \frac{e^2}{r_{ij}} |\psi_i(r_i)|^2 \, |\psi_j(r_j)|^2 \, d\tau_i d\tau_j$$

where the second term corrects for the double counting of the Coulomb repulsion of each pair of electrons. The result is not an exact one because the method takes no account of correlations between the positions of the electrons which tend to avoid one another as far as possible so as to minimize the energy.

We saw earlier the usefulness of the quantity $4\pi r^2 \, |\psi|^2$ for picturing the electron charge distribution corresponding to an atomic orbital. For the many-electron atom the above type of calculation makes possible the construction of a total charge density function by summing $4\pi r^2 \, |\psi_i|^2$ over all occupied states of an atom; an example is given in Fig. 1.5.

The calculated charge density can be compared with experiment, for it is very directly linked to the dependence on scattering vector of the atomic scattering factor for X-rays. Details will be found in the companion text in this series by Brown and Forsyth.

We can now summarize the electronic behaviour of those atoms with just one more electron than the inert rare gases, that is the alkali metals, Li, Na, K, Rb and Cs:

(i)   When electron interactions are taken into account, the electron states do divide self-consistently into closed shells, and a single outermost state. The occupancy of these states is listed in the *configuration*.

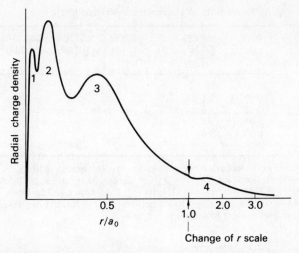

Figure 1.5   The total charge density function for the Rb$^+$ ion. The maxima are labelled with the principal quantum number of the wavefunctions making the main contributions to them.

(ii)   The outermost, valence, electron is screened from the nucleus by the closed-shell electrons. The ion core of nucleus plus closed-shell electrons is spherically symmetric, and to a first approximation behaves as a single positive charge.

(iii) The ion core electrons, because they are less well screened from the nucleus, are considerably more tightly bound than the valence electron (as can be seen in Table 1.1). For this reason, in most situations in molecules and solids they are inert, and we shall have little more to say about them.

(iv)   Because of the spherical symmetry of the ion core, angular momentum quantum numbers are still good ones for the valence electron. The radial quantum number also retains a certain significance.

The self-consistent field method of calculating one-electron energies allows us to ask such questions as: "If we add one electron to the ion core of K$^+$ ($Z = 19$, $1s^2\ 2s^2\ 2p^6\ 3s^2\ 3p^6$) will it occupy the 3d- or the 4s-orbital?" Chemistry, of course, told us the answer many years ago, but it is gratifying that physics can justify it. For all elements immediately following one with an electronic configuration $1s^2 \ldots ns^2\ np^6$ the nd one-electron level lies at higher energy than the $(n + 1)$s-level. The appearance of the periodic table* shows the expansion from short periods (8 elements) to long periods (18 elements) that accompanies the occupation by outer electrons of the $l = 2$ subshell (the d-subshell) and leads to the existence of the *transition metal* atoms. In Na and K there are

*See p. xii.

Table 1.1   Energy differences (in eV) between one-electron levels.

Atom and Z Ground state configuration	He(2) $1s^2$	Na(11) $1s^2 2s^2 2p^6 3s$	K(19) [ ]4s	Ca(20) [ ]4s²	Sc(21) [ ]3d4s²	Zn(30) [ ]3d¹⁰4s²
Excitation						
3s → 3p	0·29	2·1				
3s → 3d	0·38	3·6				
3s → 4s	0·88	3·2				
3s → 4p	0·99	3·8				
3d → 4p	0·6	0·2	−1·0	−0·6	5·5	10

The values given for the excitation energies are the smallest energy differences between appropriate configurations: e.g. $1s^1 3s^1 \rightarrow 1s^1 3p^1$ for He, 3s → 3p; $1s^2 2s^2 2p^6 3d \rightarrow 1s^2 2s^2 2p^6 4p$ for Na, 3d → 4p; *minus* [ ] 4s4p → [ ] 3d4s for Ca, 3d → 4p.
Note reversal of 3d and 4s relative positions between He and Na, and movement first up then down of 3d level relative to 4p as a function of Z.

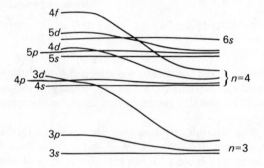

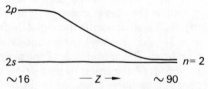

Figure 1.6   Schematic indication of *relative* positions of one-electron levels as a function of atomic number. The $ns - (n + 1)s$ separation in reality decreases with Z.

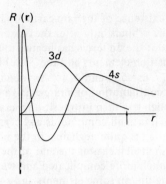

Figure 1.7   The radial parts of the 3d and 4s wavefunctions in the Fe atom. The arrow indicates a value of r equal to half the interatomic distance in metallic Fe.

large differences in the one-electron energies of the 3s-, the 3p- and the 3d-levels because of the greater efficiency of screening of the nuclear charge by the core electrons for the $l = 1$ and $l = 2$ states of a given n (the amplitudes of their wavefunctions falling to zero at the nucleus) as compared with the $l = 0$ state of the same n which has finite amplitude at the nucleus. It is important to realize however that this difference in one-electron energy levels is a strong function of the atomic number, and in Table 1.1 we give the excitation energies for the transitions 3s → 3p, 3s → 3d, 3s → 4s, 3s → 4p and 4p → 3d for certain atoms. One can thus make schematic representation of the *relative* positions of one-electron energy levels of the form shown in Fig. 1.6.

A detailed examination of the ground state configurations of transition metal atoms shows however that one cannot simply say that for this or that element the nd-levels lie below the $(n + 1)$s-level or vice versa. A ground state configuration nd^x $(n + 1)$s^1 means that a finite energy excitation is required to yield the configuration nd^{x-1} $(n + 1)$s^2 *or* that of nd^{x+1}.

Once the set of subshells ns^2 np^6 nd^{10} is complete the one-electron energies of these levels (now part of the ion core) draw together fairly rapidly as indicated by Fig. 1.6. This is understandable since the radial extent of an $n = 3$ wavefunction shrinks with increasing nuclear charge in a neutral atom, the screening effect of the extra $n > 3$ electrons being much weaker.

One other point that should be made about the one-electron states of transition metal atoms concerns the radial extent of the 3d- and 4s-wavefunctions. The presence in the radial part of the wavefunction of the terms $\exp(-r/3a_0)$ and $\exp(-r/4a_0)$ respectively means that if two iron atoms, say, are brought to the interatomic distance in iron metal the 4s–4s wavefunction overlap will be considerable, but the 3d–3d overlap much less (see Fig. 1.7) and we therefore expect 3d-wavefunctions in iron metal to be much less strongly modified from atomic wavefunctions than the 4s-functions will be.

An explanation for the existence of the rare-earth group of elements in terms of the filling up of 4f atomic orbitals only after the 5s-, 5p- and 6s-levels have been filled, and occupation of the 5d-levels has begun, follows along the lines outlined above for the transition elements and the 3d-orbitals. Configurations of the type $4f^x\,6s^2\,5d^1$, $4f^{x-1}\,6s^2\,5d^2$ and $4f^{x+1}\,6s^2$ tend to have rather similar energies and the particular configuration of a rare-earth atom depends on its environment so that (although less than in transition metals) variable valencies occur. The part-filled 4f-shell is in all compounds and crystals very much part of the ion core and 4f—4f overlap is quite negligible in the rare-earth metals.

For the partly filled 5f-shell elements at the end of the periodic table (the actinides) the situation is much more complicated and we shall not attempt to indicate it apart from saying that in some elements and compounds actinides behave more like transition metals, and in others more like rare earths.

1.4   The many-electron states of atoms

When there is more than one electron outside a closed shell, the configurational listing is still useful. So, for example, the elements that succeed Na have the following configurations in the ground state:

Na	$(1s^2)(2s^2)(2p^6)$	$3s^1$
Mg		$(3s^2)$
Al		$(3s^2)\,3p^1$
Si		$(3s^2)\,3p^2$
P		$(3s^2)\,3p^3$
S		$(3s^2)\,3p^4$
Cl		$(3s^2)\,3p^5$
Ar		$(3s^2)(3p^6)$

The energy required to permute electrons between part-filled shells of nearly the same energy is not too great, for example for Si the excited configuration $3s^1\,3p^3$ has energy 6 eV greater than the ground state configuration† (similarly, for the transition elements, to move electrons between the 3d- and 4s-shells costs little energy.)

Also, because the s- and p-shells have similar energy, the 3s-electrons are often valence electrons too, so that Al is generally trivalent in chemical combination.

What do the wavefunctions look like? First we must define what is meant by a many-particle wavefunction. Denote the electron coordinates as $r_1, r_2, r_3 \ldots$, then the many-particle wavefunction $\Psi(r_1, r_2, r_3 \ldots)$ has the property that $|\Psi(r_1, r_2, r_3 \ldots)|^2\ d\tau_1\ d\tau_2\ d\tau_3\ \ldots$ measures the probability that

---

†In fact, the exact ground state will contain a quantum mechanical admixture of these excited configurations (but only those having the same angular momentum as the initial configuration). The effects are serious only with the elements of high atomic number, where it is often impossible to identify a single dominant configuration; these are then described as having mixed configurations.

simultaneously the first electron will be found within a volume element $d\tau_1$ around the point r_1, the second at r_2, and so on.

So we might expect the two-particle wavefunction for the valence electrons in Mg (we can safely ignore the core electrons because of their inertness) to look like:

$$\Psi(r_1, r_2) = \psi_{3s}(r_1)\psi_{3s}(r_2)$$

where $\psi_{3s}(r)$ is a self-consistent 3s-wavefunction in the field of a Mg nucleus and eleven other electrons. Notice that there is no spatial correlation whatsoever between the two 3s-electrons, for

$$|\Psi(r_1, r_2)|^2 = |\psi_{3s}(r_1)|^2 |\psi_{3s}(r_2)|^2$$

and the probability distributions of the electrons are independent: also the charge distribution is spherically symmetric.

We have also to specify the spin arrangement; because of the exclusion principle, the electrons must have opposite spin, say $(\uparrow_1, \downarrow_2)$, or $(\downarrow_1, \uparrow_2)$ would be equally good. Because electrons are indistinguishable, both combinations must be included, so as to form either

$$[(\uparrow_1, \downarrow_2) + (\downarrow_1, \uparrow_2)]/\sqrt{2} \quad \text{spin symmetric}$$

or

$$[(\uparrow_1, \downarrow_2) - (\downarrow_1, \uparrow_2)]/\sqrt{2} \quad \text{spin antisymmetric}$$

where the symmetry or antisymmetry refers to what happens when the electron labels 1 and 2 are interchanged. A cardinal principle of quantum mechanics is: for electrons the *total* wavefunction, spin and spatial parts combined, must be antisymmetric, and change sign when any pair of electron labels are interchanged. Consequently the total wavefunction for the two 3s-electrons is

$$\Psi = \psi_{3s}(r_1)\psi_{3s}(r_2)[(\uparrow_1, \downarrow_2) - (\downarrow_1, \uparrow_2)]/\sqrt{2}$$

When a third valence electron is added (Al) we might ask which of the six available 3p-orbitals it goes into, but for a free atom the question is not meaningful. The reason is that those orbitals differ only in the direction, not in magnitude, of orbital and spin angular momentum; but there is no reference direction associated with the core plus two 3s-electrons, since that charge density is spherically symmetric. Consequently, in the absence of an external field, the electron states are completely specified† by the configuration. The angular momenta of the atom are just those of the single 3p-electron. The total orbital and spin angular momenta of an atom are denoted by capital letters L and S, and their components as M_L and M_S. Just as with a single electron, the compact spectroscopic notation is still in use to describe the values of L and S. A many-electron state with total orbital angular momentum magnitude $0,1,2,3,4\ldots\hbar$ is denoted as S,P,D,F,G, $\ldots$; the total spin angular momentum

†Apart from the relative direction of the orbital and spin angular momentum—see p 16.

magnitude is written as a superscript in the form $(2S + 1)$ (the reason for this is that $(2S + 1)$ is the number of different values of M_S). The ground state of Al is therefore denoted as ^{2}P, and spoken of as 'doublet P'; the ground state of Mg is ^{1}S ('singlet S'), and Na ^{2}S ('doublet S').

With the addition of another 3p-electron, the question of which p-orbitals are occupied *is* significant, for the aspherical wavefunction of one of them provides a reference direction for the other. Bearing in mind the exclusion principle, there are $6 \times 5/2$ possible pairs of occupied orbitals (because electrons are indistinguishable it is meaningless to ask which of the two electrons is in which state).

The rules for addition of orbital angular momenta are that the maximum value of L is given by the maximum value of $| \Sigma_i m_{l_i} |$ over all the orbitals that are occupied subject to the constraints of the exclusion principle, and the minimum value of L is the minimum of $| \Sigma_i m_{l_i} |$; L then takes all integer values between these minimum and maximum values. The allowed values of S are obtained similarly. For example, with two 3p-electrons, the maximum value of L comes from arranging the two electrons to have parallel orbital angular momenta, say $m_{l_1} = m_{l_2} = +1$ (or equally $m_{l_1} = m_{l_2} = -1$); because of the exclusion principle the electron spins must be antiparallel, $S = 0$. The minimum value of L comes from pairs such as $m_{l_1} = 1, m_{l_2} = -1$, and is equal to zero. Consequently the allowed values of L will be 2,1, and 0. The minimum and maximum values of S are 0 and 1, clearly there are no other allowed integer values.

What do the wavefunctions look like? In general, a simple product wavefunction $\psi_{3p}^{+1}(r_1)\psi_{3p}^{-1}(r_2)[\uparrow_1, \uparrow_2]$ (the superscripts denote the value of m_l) is not a good eigenstate. The reason lies with the spatial part of the wavefunction. Because of electron–electron repulsion one p-electron creates an asymmetric potential for the other, and vice versa, consequently the orbital angular momentum of each electron considered separately is *not* a good quantum number. However, if we segregate the two p-electrons from the rest of the system, the latter *does* provide a spherically symmetric potential, so although m_{l_1} and m_{l_2} are not good quantum numbers, the total orbital angular momentum $M_L(=m_{l_1} + m_{l_2})$ is conserved and does provide a good quantum number. We have therefore to seek eigenstates of M_L, and also of M_S.

The eigenstates (see Appendix A.1.2, p. 19) group into three *terms* that have some orbital or spin degeneracy:

^{1}D	$L = 2$,	$M_L = 2,1,0,-1,-2$	$S = 0, M_S = 0$	5 states
^{3}P	$L = 1$,	$M_L = 1,0,-1$	$S = 1, M_S = 1,0,-1$	9 states
^{1}S	$L = 0$,	$M_L = 0$	$S = 0, M_S = 0$	1 state

Within each term states having different values of M_L or M_S are identical in energy unless there is an external field, for they differ only in the direction of their angular momenta†. There are altogether 15 of these two-electron states, that are made up from 15 linearly independent combinations of the pairs of states discussed earlier; the conservation of the number of states is a characteristic feature of quantum mechanics.

The probability distributions of the two electrons are no longer independent, and there is now some correlation between them. For example, one of the states in the ^{3}P term (because they all must have the same energy, it does not matter which of the nine states we consider) has two-particle wavefunction

$$\Psi_{M_L=1, \, M_S=1}(r_1, r_2) = 2^{-1/2}(\psi_{3p}^{+1}(r_1)\psi_{3p}^0(r_2) - \psi_{3p}^0(r_1)\psi_{3p}^{+1}(r_2))[\uparrow_1, \uparrow_2]$$

which is identically zero if $r_1 = r_2$. The antisymmetrization of the two-particle wavefunction ensures that the electrons are kept apart (see Fig. A.1.2.1); to put it more simply still, electrons of parallel spin cannot have the same spatial coordinates. On the other hand, the wavefunctions of both the ^{1}D and ^{1}S terms do allow the two electrons to occupy the same part of space, because the two electrons have opposite spin. The average electron–electron repulsion will therefore be less for the ^{3}P term than for the ^{1}D and ^{1}S terms, and since the other energies are the same for all three terms, the ground state term will be the ^{3}P one. In atomic Si, this is indeed found to be ground state, and the ^{1}D and ^{1}S terms are 0.8 eV and 1.9 eV higher in energy.

We can now summarize the situation for atoms with several electrons outside a closed shell:

(i)   Self-consistent field one-electron states can be found, which are analogous to the shells of hydrogenic atoms and the occupancy of these is listed in the configuration.

(ii) The individual one-electron quantum numbers m_l and m_s no longer hold good; instead it is the total angular momenta M_L and M_S that are good quantum numbers.

(iii) The many-electron states are grouped into terms, each characterized by a particular value of L and S. The allowed combinations of spin and orbital wavefunctions are determined by the antisymmetry requirement on the total wavefunction.

(iv) In the absence of an external field (and before taking spin–orbit coupling into account) all the $(2L + 1)(2S + 1)$ states within a term have the same energy. Terms themselves differ markedly in energy, because they represent different electron–electron correlations, and so different average electron–electron repulsive energy.

Empirically, it is a fairly general rule (the first of Hund's rules in spectroscopy) that the lowest-lying term is the one of maximum spin ('multiplicity' in

†We are still ignoring the spin–orbit interaction energy, see p 16.

spectroscopists' language). The reason is that in that case as many electrons as possible have like spin, and are therefore forced to keep away from each other by the exclusion principle. If more than one term has the maximum value of S, Hund's second rule states that it is the term of maximum L that lies lowest; this too corresponds to favouring states in which the many-electron wavefunction keeps the electrons as far apart as possible.

These ideas are, because they depend only on symmetry arguments, of quite general application; they provide a means of characterizing the eigenstates of complicated many-electron atoms, and an estimate of the order in which the energy levels lie. It is a very much more difficult matter to calculate the energies themselves, and not something that we shall attempt. For our purposes of considering what happens with aggregates of atoms, it is more useful to take the energies from extensive tabulations obtained from atomic spectra.

Finally let us see how to apply the ideas to some examples:

(i)   Atomic P. Ground state configuration $(1s^2)(2s^2 2p^6)3s^2 3p^3$. Maximum allowed $S = 3/2$, obtained by occupying orbitals with $m_l = 1, 0$ and -1 with electrons of like spin. Ground state term is ^4S.

(ii)   Atomic V. Ground state configuration $(1s^2)(2s^2 2p^6)(3s^2 3p^6)3d^3 4s^2$. There are five 3d-orbitals, having $m_l = 2, 1, 0, -1, -2$, and the maximum S is given by singly occupying any three of them. With this value of S, the maximum L is 3 and the minimum is 0. Ground state term is ^4F.

We have not yet discussed effects which follow from the interaction between an atom's orbital angular momentum and its spin. Consideration of such effects needs a new quantum number which specifies the total angular momentum. For a single electron this is j (with z-component m_j) and for a whole atom J with z-component M_J. Thus a hydrogen atom in the excited $2p^1$ configuration (^2P) can have $J = L + S$ or $L-S$ and these two states $^2P_{3/2}$ and $^2P_{1/2}$ have different energies because of the spin–orbit coupling $\lambda \mathbf{L} \cdot \mathbf{S}$. The rules governing the combinations of M_L and M_S to give M_J are like those governing the combinations of m_{l_1} and m_{l_2} to give M_L for two electrons, and one may group together various M_J values to specify states of different J as one grouped values of M_L and M_S to give terms of given L and S. With spin–orbit interaction taken into account, the state of the atom is defined by its configuration, and the values of L, S, J and M_J. None of M_L, M_S, m_l or m_s are, in general, good quantum numbers.

States of given L and S but different J are separated in energy by the spin-orbit coupling, but these separations are much smaller than those of different terms where large   electron–electron Coulomb interactions were involved. A simple empirical rule governing the order of the J states is that for a less than half-filled p-, d- or f-shell that of lowest energy has $J = L - S$ with a ladder of states up to $J = L + S$, while for a more than half-filled shell the reverse situation holds.

It might seem at first sight that considerations of these angular momentum states of partly-filled shells are irrelevant in molecules or solids since the electrons in such shells are those involved in the bonding process, and they are subject to a more complicated potential. It must be recalled, though, that in salts of transition metals (e.g. $FeCl_2$) and in salts, alloys, and elemental metals of rare-earth elements there exist partly filled d- or f-shells which are not involved in bonding and have a net angular momentum and so also a magnetic moment. Thus Er^{+++} in Er metal, $ErCl_3$ or a Au–Er alloy has a configuration $4f^{11}$ with $S = 3/2$ $L = 6$ as the ground state, represented $^4I_{15/2}$ because $J = L + S$ for the lowest term of this more than half-full shell.

A.1.1   Appendix: Hydrogenic wavefunctions for $n = 1$ to $n = 3$

$$\psi_{1s}^0 = \left(\frac{Z^3}{\pi a_0}\right)^{1/2} e^{-Zr/a_0}$$

$$\psi_{2s}^0 = \frac{1}{4\sqrt{2}} \left(\frac{Z^3}{\pi a_0}\right)^{1/2} (2 - Zr/a_0)e^{-Zr/2a_0}$$

$$\psi_{2p}^0 = \frac{1}{4\sqrt{2}} \left(\frac{Z^3}{\pi a_0}\right)^{1/2} (Zr/a_0)e^{-Zr/2a_0} \cos\theta$$

$$\psi_{2p}^{\pm 1} = \frac{1}{8} \left(\frac{Z^3}{\pi a_0}\right)^{1/2} (Zr/a_0)e^{-Zr/2a_0} \sin\theta\, e^{\pm i\phi}$$

$$\psi_{3s}^0 = \frac{1}{81\sqrt{3}} \left(\frac{Z^3}{\pi a_0}\right)^{1/2} (27 - 18Zr/a_0 + 2Z^2r^2/a_0^2)e^{-Zr/3a_0}$$

$$\psi_{3p}^0 = \frac{\sqrt{2}}{81} \left(\frac{Z^3}{\pi a_0}\right)^{1/2} (6 - Zr/a_0)(Zr/a_0)e^{-Zr/3a_0} \cos\theta$$

$$\psi_{3p}^{\pm 1} = \frac{1}{81} \left(\frac{Z^3}{\pi a_0}\right)^{1/2} (6 - Zr/a_0)(Zr/a_0)e^{-Zr/3a_0} \sin\theta\, e^{\pm i\phi}$$

$$\psi_{3d}^0 = \frac{1}{81\sqrt{6}} \left(\frac{Z^3}{\pi a_0}\right)^{1/2} (Zr/a_0)^2 e^{-Zr/3a_0} (3\cos^2\theta - 1)$$

$$\psi_{3d}^{\pm 1} = \frac{1}{81} \left(\frac{Z^3}{\pi a_0}\right)^{1/2} (Zr/a_0)^2 e^{-Zr/3a_0} \sin\theta \cos\theta\, e^{\pm i\phi}$$

$$\psi_{3d}^{\pm 2} = \frac{1}{162} \left(\frac{Z^3}{\pi a_0}\right)^{1/2} (Zr/a_0)^2 e^{-Zr/3a_0} \sin^2\theta\, e^{\pm 2i\phi}$$

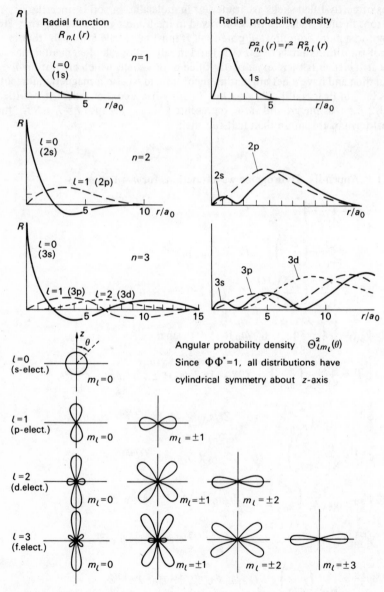

Figure A.1.1.1   The shapes of the hydrogenic wavefunctions and charge densities for $n = 1$ to $n = 3$.

A.1.2   Appendix: The states of a two-electron atom

In Section 1.4 we saw that when electron–electron interactions are taken into account, the individual orbital angular momenta m_l are no longer good quantum numbers. However, the total orbital angular momentum M_L is conserved (and incidentally also the total spin M_S). We have therefore to look for a two-particle wavefunction.

$$\Psi_{LM_L,\,SM_S}(r_1,r_2) \qquad (\equiv \Psi_{L,\,M_L}^{\text{spatial}}(r_1,r_2)\Psi_{S,\,M_S}^{\text{spin}}(1,2))$$

which is an eigenstate of the operator $\hat{M}_L$ (but not, in general, of $\hat{m}_{l_1}$ and $\hat{m}_{l_2}$). We assume that the total wavefunction, which must be antisymmetric with respect to interchange of electron labels, is a simple product of spatial and spin parts; this assumption is equivalent to neglect of spin–orbit interaction.

$\hat{M}_L$ can be expressed in terms of $\hat{m}_{l_1}$ and $\hat{m}_{l_2}$ treating the operators as if they were classical vectors:

$$\hat{M}_L = \hat{m}_{l_1} + \hat{m}_{l_2}$$

Similarly, the operator for the total square magnitude of angular momentum is:

$$\hat{L}^2 = (\hat{l}_1 + \hat{l}_2)^2 = \hat{l}_1^2 + 2\hat{l}_1.\hat{l}_2 + \hat{l}_2^2$$

$2(\hat{l}_1.\hat{l}_2)$ can most usefully be written as

$$(\hat{l}_{1+}\hat{l}_{2-} + \hat{l}_{1-}\hat{l}_{2+} + 2\hat{m}_{l_1}\hat{m}_{l_2})$$

where $\hat{l}_{1\pm} \equiv \hat{l}_{x1} \pm i\hat{l}_{y1}$; $\hat{m}_{l_1}$ is of course our notation for $\hat{l}_{z1}$.

The one-electron states that we were concerned with in Section 1.4 were those of the 3p-shell, $\psi_{3p}^{+1}(r)$, $\psi_{3p}^{0}(r)$ and $\psi_{3p}^{-1}(r)$, where the superscripts denote the value of m_l. For the present problem the value of m_l alone is sufficient to specify these states, and n and l are essentially redundant. We can therefore use a more compact notation for the three orbitals $|+1\rangle$ for ψ_{3p}^{+1}, $|0\rangle$ for ψ_{3p}^{0} and $|-1\rangle$ for ψ_{3p}^{-1}.

The rules for the operators are:

$$\hat{m}_l\,|\,m_l\,\rangle = m_l\,|\,m_l\,\rangle$$
$$\hat{l}^2\,|\,m_l\,\rangle = l(l+1)\,|\,m_l\,\rangle \quad (l=1 \text{ for p states})$$
$$\hat{l}_\pm\,|\,m_l\,\rangle = [(l \mp m_l)(l \pm m_l + 1)]^{1/2}\,|\,m_l \pm 1\,\rangle$$

The $\hat{l}_\pm$ operators are, for obvious reasons, called the raising and lowering operators; notice that $\hat{l}_+\,|\,(m_l = 1)\,\rangle = \hat{l}_-\,|\,(m_l = -1)\,\rangle = 0$, so that these operators correctly terminate, and allow m_l to run from $+1$ to -1 only.

An operator operates only on its own electron, e.g.

$$\hat{m}_{l_2}\,|\,m_{l_1}\,\rangle\,|\,m_{l_2}\,\rangle \equiv |\,m_{l_1}\,\rangle\,\hat{m}_{l_2}\,|\,m_{l_2}\,\rangle = m_{l_2}\,|\,m_{l_1}\,\rangle\,|\,m_{l_2}\,\rangle$$

We are now ready to set up the two-electron states. Suppose, as suggested in Section 1.4, we try out $\Psi^{\text{spatial}} = (|+1\rangle |+1\rangle)$; is it an eigenstate, and what quantum numbers does it have (the first term represents the state of electron 1 and the second of electron 2)?

$$\hat{M}_L(|+1\rangle |+1\rangle) = (\hat{m}_{l_1} + \hat{m}_{l_2})(|+1\rangle |+1\rangle)$$

$$= (+|+1\rangle |+1\rangle + |+1\rangle |+1\rangle) = 2(|+1\rangle |+1\rangle)$$

therefore this state is indeed an eigenstate of $\hat{M}_L$, with eigenvalue M_L equal to 2. Consider now

$$\hat{L}^2(|+1\rangle |+1\rangle) = [\hat{l}_1^2 + (\hat{l}_{1+}\hat{l}_{2-} + \hat{l}_{1-}\hat{l}_{2+} + 2\hat{m}_{l_1}\hat{m}_{l_2}) + \hat{l}_2^2] \, (|+1\rangle |+1\rangle)$$

We have

$$\hat{l}_1^2(|+1\rangle |+1\rangle) = \hat{l}_2^2(|+1\rangle |+1\rangle) = l(l+1)(|+1\rangle |+1\rangle)$$

$$\hat{m}_{l_1}\hat{m}_{l_2}(|+1\rangle +1\rangle) = (+1)(+1)(|+1\rangle |+1\rangle)$$

$$\hat{l}_{+1}\hat{l}_{2-}(|+1\rangle |+1\rangle) = \hat{l}_{1-}\hat{l}_{2+}(|+1\rangle |+1\rangle) = 0$$

therefore

$$\hat{L}^2(|+1\rangle |+1\rangle) = 2[l(l+1) + 1](|+1\rangle |+1\rangle)$$

Since we are dealing with p-electrons (and notice that the l quantum number is still a good one), $l = 1$, so that the eigenvalue is 6. In the usual manner we write

$$\hat{L}^2(|+1\rangle |+1\rangle) = L(L+1)(|+1\rangle |+1\rangle)$$

and we have shown L to be equal to 2, that is, a D-state.

We have now found the two-electron state $\Psi^{\text{spatial}}_{L, M_L}$ with quantum numbers L, M_L equal to 2 and 2; in the same manifold there must be a further four states with $M_L = 1, 0, -1, -2$. The easiest way to find them is to operate with

$$\hat{L}_- \Psi^{\text{spatial}}_{L, M_L} = [(L + M_L)(L - M_L + 1)]^{1/2} \Psi^{\text{spatial}}_{L, M_L - 1}$$

which is the many-electron analogue of the one-electron lowering operator:

$$\hat{L}_- \Psi^{\text{spatial}}_{2,2} = (4.1)^{1/2} \Psi^{\text{spatial}}_{2,1}$$

$$\hat{L}_- \Psi^{\text{spatial}}_{2,2} \equiv (\hat{l}_{1-} + \hat{l}_{2-})(|+1\rangle |+1\rangle)$$

$$= [2.1]^{1/2} |0\rangle |+1\rangle + [2.1]^{1/2} |+1\rangle |0\rangle$$

Therefore

$$\Psi^{\text{spatial}}_{2,1} = 2^{-1/2}(|0\rangle |+1\rangle + |+1\rangle |0\rangle)$$

Successive applications of $\hat{L}_-$ gives the other three D-states (Table A.1.2.1).

Table A.1.2.1 Two-electron total wavefunctions for the p^2 configuration. $\Psi_{L, M_L, S, M_S} = \Psi^{spatial}_{L, M_L} \Psi^{spin}_{S, M_S}$.

Term	$\Psi^{spatial}_{L, M_L}$	L	M_L	Ψ^{spin}_{S, M_S}	S	M_S
¹D 5 states	$|+1\rangle |+1\rangle$	2	2			
	$2^{-1/2}(|+1\rangle|0\rangle + |0\rangle|+1\rangle)$	2	1			
	$6^{-1/2}(|+1\rangle|-1\rangle + 2|0\rangle|0\rangle + |-1\rangle|+1\rangle)$	2	0	$2^{-1/2}(|\uparrow\rangle|\downarrow\rangle - |\downarrow\rangle|\uparrow\rangle)$	0	0
	$2^{-1/2}(|-1\rangle|0\rangle + |0\rangle|-1\rangle)$	2	-1			
	$|-1\rangle|-1\rangle$	2	-2			
³P 9 states	$2^{-1/2}(|+1\rangle|0\rangle - |0\rangle|+1\rangle)$	1	1	$|\uparrow\rangle|\uparrow\rangle$	1	1
	$2^{-1/2}(|+1\rangle|-1\rangle - |-1\rangle|+1\rangle)$	1	0	$2^{-1/2}(|\uparrow\rangle|\downarrow\rangle + |\downarrow\rangle|\uparrow\rangle)$	1	0
	$2^{-1/2}(|-1\rangle|0\rangle - |0\rangle|-1\rangle)$	1	-1	$|\downarrow\rangle|\downarrow\rangle$	1	-1
¹S	$3^{-1/2}(|+1\rangle|-1\rangle - |0\rangle|0\rangle + |-1\rangle|+1\rangle)$	0	0	$2^{-1/2}(|\uparrow\rangle|\downarrow\rangle - |\downarrow\rangle|\uparrow\rangle)$	0	0

What about the P-states? Suppose we look for $\Psi_{1,1}^{spatial}$, we have to find a state that, like $\Psi_{2,1}^{spatial}$, has $M_L = 1$, but it must be orthogonal to $\Psi_{2,1}$. This suggests that we try:

$$\Psi_{1,1}^{spatial} = 2^{-1/2}(|\,0\,\rangle\,|+1\,\rangle - |+1\,\rangle\,|\,0\,\rangle)$$

Clearly $\hat{M}_L \Psi_{1,1}^{spatial} = 1 \cdot \Psi_{1,1}^{spatial}$, therefore $M_L = 1$.

$$\hat{L}^2 \Psi_{1,1}^{spatial} = [\hat{l}_1^2 + \hat{l}_2^2 + 2\hat{m}_{l_1}\hat{m}_{l_2} + (\hat{l}_{1+}\hat{l}_{2-} + \hat{l}_{1-}\hat{l}_{2+})]\,\Psi_{1,1}^{spatial}$$
$$= (2 + 2 + 2.0)\Psi_{1,1}^{spatial} + 2^{-1/2}\{(1.2)^{1/2}(2.1)^{1/2}|+1\,\rangle\,|\,0\,\rangle$$
$$+ 0 + 0 - (2.1)^{1/2}(1.2)^{1/2}\,|\,0\,\rangle\,|+1\,\rangle\}$$
$$= (2 + 2 - 2)\,\Psi_{1,1}^{spatial} = 2\Psi_{1,1}^{spatial}$$

Therefore $L = 1$(P-state) as required.

The two other P-states are found by applying $\hat{L}_-$ to $\Psi_{1,1}^{spatial}$.

The one remaining state is $\Psi_{0,0}^{spatial}$. To ensure that $\hat{M}_L$ has eigenvalue zero, it must contain only the pairs $(|+1\,\rangle\,|-1\,\rangle)$, $(|\,0\,\rangle\,|\,0\,\rangle)$ and $(|-1\,\rangle\,|+1\,\rangle)$. Furthermore, it must be orthogonal to $\Psi_{2,0}^{spatial}$ and $\Psi_{1,0}^{spatial}$; the correct linear combination is given in Table A.1.2.1. Application of the $\hat{L}^2$ operator does, after tedious manipulation, show that $L = 0$ for this state (S-state).

We now wish to find Ψ^{spin}. For both D- and S-terms $\Psi^{spatial}$ is symmetric with respect to interchange of electron labels, consequently Ψ^{spin} must be antisymmetric. The only combination of pairs that is antisymmetric is

$$\Psi_{0,0}^{spin} = 2^{-1/2}(|\uparrow\rangle\,|\downarrow\rangle - |\downarrow\rangle\,|\uparrow\rangle)$$

We have indicated that this spin state has quantum numbers $S = 0$, $M_S = 0$ which is easily proved (the rules for spin operators parallel those for orbital operators):

$$\hat{S}^2 \Psi_{0,0}^{spin} = [\hat{s}_1^2 + \hat{s}_2^2 + 2\hat{m}_{s_1}\hat{m}_{s_2} + (\hat{s}_{1+}\hat{s}_{2-} + \hat{s}_{1-}\hat{s}_{2+})]\,\Psi_{0,0}^{spin}$$
$$= (\tfrac{1}{2}\cdot\tfrac{3}{2} + \tfrac{1}{2}\cdot\tfrac{3}{2} - 2\cdot\tfrac{1}{2}\cdot\tfrac{1}{2})\Psi_{0,0}^{spin} + [0 - 2^{-1/2}(1\cdot1)^{1/2}|\uparrow\rangle\,|\downarrow\rangle$$
$$+ 2^{-1/2}(1\cdot1)^{1/2}|\downarrow\rangle\,|\uparrow\rangle + 0]$$
$$= (\tfrac{3}{4} + \tfrac{3}{4} - \tfrac{1}{2} - 1)\Psi_{0,0}^{spin} = 0$$

$$\hat{M}_S \Psi_{0,0}^{spin} = (\hat{m}_{s_1} + \hat{m}_{s_2})\Psi_{0,0}^{spin}$$
$$= 2^{1/2}\cdot\tfrac{1}{2}\cdot(|\uparrow\rangle\,|\downarrow\rangle + |\downarrow\rangle\,|\uparrow\rangle - |\uparrow\rangle\,|\downarrow\rangle - |\downarrow\rangle\,|\uparrow\rangle) = 0$$

One spin symmetric state is

$$\Psi_{1,1}^{spin} = |\uparrow\rangle\,|\uparrow\rangle$$
$$\hat{M}_s \Psi_{1,1}^{spin} = (\tfrac{1}{2} + \tfrac{1}{2})\Psi_{1,1}^{spin} = 1 \cdot \Psi_{1,1}^{spin}; \quad M_s = 1$$

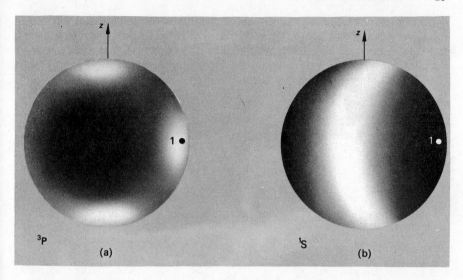

Figure A.1.2.1   Electron—electron correlation in the p² problem. The probability density of electron 2 for electron 1 at the position shown (a) for the $M_L = 0$ state of the ^{3}P term (b) for the ^{1}S term.

as required. Also,

$$\hat{S}^2 \Psi_{1,1}^{spin} = [\tfrac{1}{2} \cdot \tfrac{3}{2} + \tfrac{1}{2} \cdot \tfrac{3}{2} + 2 \cdot \tfrac{1}{2} \cdot \tfrac{1}{2} \cdot + (0 + 0)] \Psi_{1,1}^{spin}$$

$$= 2\Psi_{1,1}^{spin} \quad \text{therefore } S = 1.$$

The other states, with $M_s = 0, -1$, can be written down immediately.

We have dealt with the simplest non-trivial many-electron states, and the procedure for dealing with more electrons, or electrons of greater angular momenta, follows the same general principles. The allowed terms can be found easily by application of the rules given in Section 1.4; the calculation of the many-electron states, as has been done in this Appendix, always involves much tedious algebra, but it can be somewhat eased by the application of group theory. The total wavefunction can always be written as a linear combination of products of one-electron wavefunctions, as in Table A.1.2.1, and the coefficients (known as Clebsch—Gordon coefficients) are tabulated in standard texts for not too large values of L and M_L.

The importance of setting up the correct many-electron wavefunction is that it enables the electron—electron repulsion to be calculated properly;

indirectly it is the symmetry constraints on the total wavefunction that control the correlations of electrons with each other. The Hund rules then can be shown to follow naturally from the relative amounts of electron—electron repulsion in the different terms of a configuration. Diagrams showing schematically the charge clouds that enable one to visualize this repulsion are shown in Fig. A.1.2.1.

2

Bonding Between Atoms

2.1 Introduction

When we form an elemental solid by condensation from a gas of weakly interacting atoms the bonding energy arises from modifications in the states of the outer electrons, such that they are accommodated at lower energies. In simple materials that are solid at room temperature these electrons are a small number (per atom) of well-defined electrons which are termed the valency electrons, since they are those that take part in chemical bonding processes with other types of atoms to form molecules; an understanding of such chemical bonding processes is clearly a valuable step towards an understanding of the electronic structures of solids. Unfortunately many students of physics nowadays have only limited awareness of chemistry and we shall not assume any detailed chemical knowledge beyond the following:

   (i)   On the left hand side of the periodic table elements in the first three
         columns have valencies equal to their group number, and tend to form
         ionic compounds in which they are positive ions with closed shells.
   (ii) Elements on the right hand side of the periodic table in groups VI and
         VII have valencies equal to 8 minus the group number, and tend to
         form ionic compounds in which they are negative ions with closed
         shells.
   (iii) Elements in the transition groups show various valencies; the outer
         s-electrons are always involved but no simple rules can be given for the
         involvement of the d-electrons in bonding.
   (iv) Elements in the rare-earth group are almost always trivalent in
         compounds with the main exceptions of Eu (often divalent) Yb
         (sometimes divalent) and Ce (sometimes 4-valent).

These features are not unreasonable in the light of the general characteristics of atomic energy levels that we discussed in the previous chapter, and often in the solid there is an important contribution to cohesion from the Coulomb attraction between charged ions of opposite sign. For example, in crystalline LiF it has cost $5 \cdot 3$ eV (the ionization energy) to remove the outer electron from Li to form Li^+, but $3 \cdot 4$ eV (the electron affinity) is gained by adding that electron to F to form F^-, and a further $10·7$ eV comes from the electrostatic attraction between the ions in the crystal; the net gain is therefore $(10·7 + 3·4 - 5·3) = 8·8$ eV per LiF. This, however, is an extreme modification in the states occupied by the outer electrons and one that is only possible when atoms of different types (and of what chemists call very different electronegativity) are involved.

In general, chemical binding can involve arbitrary mixtures of electron transfer and electron sharing between atoms, so that terms like ionic and covalent (electron-sharing) must be used with care. Since we are concerned first to give an account of the electronic structures of elemental solids we shall begin this chapter by a discussion of the states in which an electron or a pair of electrons can be accommodated at low energy in a system of two identical nuclei, specifically hydrogen nuclei. After extending the ideas developed to pairs or larger groups of more complicated atoms, we shall consider what other types of interaction need to be invoked to explain the features of bonding in such substances as inert gas crystals and ice. It should then be possible for us to understand in a general way the observed structures and binding energies of the elemental solids.

2.2   Molecular orbitals in H_2^+ and H_2

In principle we should go about the problem of electron states in molecules as we did that of electron states in atoms, solving the Schrödinger equation and calculating the one-electron energies with a potential provided by all the nuclei and all the other electrons. In practice the loss of spherical symmetry makes a rigorous procedure of this sort impossible. Fortunately the normal internuclear spacings in molecules are such that the one-electron states can be given to a first approximation by the combination of the wavefunctions describing the one-electron states of the component atoms. Thus we construct *molecular orbitals* to accommodate the electrons involved in the bonding by a linear combination of atomic orbitals (LCAO). (In a heteronuclear diatomic molecule we would have

$$\psi_{\text{m.o.}} = p\psi_a + q\psi_b$$

where ψ_a is an atomic orbital on one atom, ψ_b an atomic orbital on the other and p and q are chosen to minimize the one-electron energy; but in H_2 and H_2^+ symmetry clearly requires $|p| \equiv |q|$.) With explicit expressions for the molecular orbital wavefunctions the one-electron energies can be calculated and the total energy of the molecule compared with that of the separated atoms. (This was not historically the method used by Heitler and London in the first calculation of the covalent bonding energy of H_2, but it has the advantage over their approach (see p. 30) that it can be generalized to molecular orbitals involving more than two atoms, and allows one to envisage the electron configurations of molecules in terms of the occupation successively of a set of one-electron levels as in the many-electron atom.)

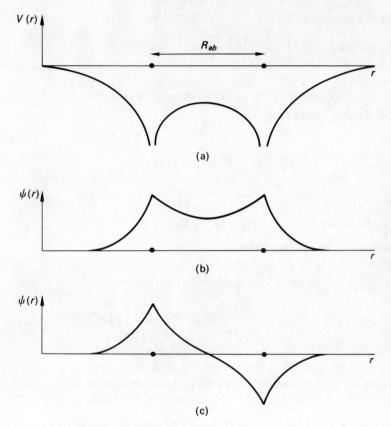

Figure 2.1 The potential (a) and possible molecular orbital wavefunctions, (b) bonding and (c) antibonding for H_2^+.

In H_2^+ then we have a potential like that of Fig. 2.1a and an unnormalized ground state wavefunction that we guess† as either (see Figs. 2.1b and 2.1c)

$$\psi_+ = \psi_a(r - R_a) + \psi_b(r - R_b)$$

or

$$\psi_- = \psi_a(r - R_a) - \psi_b(r - R_b)$$

where r is the position coordinate of the electron, R_a and R_b those of the nuclei of the two atoms a and b and ψ_a and ψ_b are 1s hydrogen atom wavefunctions

†Application of the variation method (choice of p/q that makes E least) shows that the best LCAO is in fact given by $p/q = \pm 1$.

for the two atoms, i.e.,

$$\psi_a(r - R_a) = \exp\left(-\frac{r - R_a}{a_0}\right) = \psi_a$$

in this section.

The energy of an electron in these two states $\psi_\pm$ is then given by

$$E = \int \psi^* \mathcal{H} \psi \, d\tau / \int \psi^* \psi \, d\tau$$

where

$$\mathcal{H} = -\frac{\hbar^2}{2m} \nabla^2 - \frac{e^2}{r_a} - \frac{e^2}{r_b} + \frac{e^2}{R_{ab}}$$

where

$$r_a = |r - R_a| \quad \text{and} \quad R_{ab} = |R_a - R_b|$$

Mathematical manipulation yields

$$E_\pm = \frac{e^2}{R_{ab}} + \frac{E'_a \pm \beta}{1 \pm S'}$$

where

$$E'_a = \int \psi_a^* \mathcal{H}' \psi_a \, d\tau, \qquad \beta = \int \psi_a^* \mathcal{H}' \psi_b \, d\tau = \int \psi_b^* \mathcal{H}' \psi_a \, d\tau$$

$$\mathcal{H}' = -\frac{\hbar^2}{2m} \nabla^2 - \frac{e^2}{r_a} - \frac{e^2}{r_b}, \qquad S' = \int \psi_a \psi_b \, d\tau$$

$\mathcal{H}'$ is thus a truncated Hamiltonian operator that leaves out the nuclear repulsion (which is a function of R_{ab} only), and S' is an overlap integral.

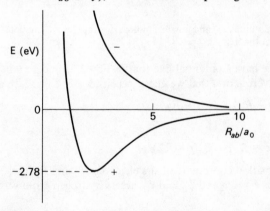

Figure 2.2   Energy as a function of internuclear separation for the bonding and antibonding molecular orbitals in H_2^+. (The zero of energy is that of $H + H^+$.)

From the definition of ψ_a (which is real) we know that

$$\left(-\frac{\hbar^2}{2m}\nabla^2 - \frac{e^2}{r_a}\right)\psi_a = E_a\psi_a$$

where E_a is the energy of the neutral a atom, hence

$$E_a' = \int \psi_a\left(E_a - \frac{e^2}{r_b}\right)\psi_a d\tau = E_a - \int \frac{e^2}{r_b}\psi_a^2 d\tau$$

similarly

$$\beta = E_a S' - \int \frac{e^2}{r_b}\psi_a\psi_b d\tau$$

so

$$E_{\pm} = E_a + \frac{e^2}{R_{ab}} - \frac{\int(e^2/r_b)\psi_a^2 d\tau \pm \int(e^2/r_b)\psi_a\psi_b d\tau}{1 \pm S'} \tag{2.1}$$

For the ψ_+ this does result in stability for the H_2^+ ion with respect to a separated H atom and H^+ ion when the nuclear separation is about 1 Å (see Fig. 2.2). For ψ_- the energy of H_2^+ is higher and this is therefore called the antibonding molecular orbital. The binding for ψ_+ can be given a simple physical interpretation: ψ_+ provides a heightened probability of finding the electron in the attractive potential region between the nuclei (see Fig. 2.1).

A similar calculation can be carried out for the H_2 molecule using the same wavefunctions, but the Hamiltonian $\mathcal{H}'$ will now include the terms

$$-\frac{e^2}{r_{a1}} - \frac{e^2}{r_{b2}} - \frac{e^2}{r_{a2}} - \frac{e^2}{r_{b1}} + \frac{e^2}{r_{1,2}}$$

where the coordinate r_{a1} is that of electron 1 with respect to atom a etc.

The one-electron energy ϵ of an electron in the ψ_+ molecular orbital will be

$$\epsilon = \int\psi_+\mathcal{H}'\psi_+ d\tau / \int\psi_+\psi_+ d\tau$$

and operating as we did for H_2^+ we find terms like E_a' and β (containing now however $e^2/r_{1,2}$ terms).

Thus $\epsilon_1 + \epsilon_2$ (if we use the ψ_+ orbital for each) will contain in place of Equ. 2.1

(i)   twice the E_a term $+27\cdot2$ eV
(ii)   twice the terms that replace the integrals in Equ. 2.1 $-32\cdot4$ eV
(iii)   *once* the nuclear repulsion $+19\cdot3$ eV

                                                               $+14\cdot1$ eV

The energy of the molecule will be less than this sum of the one-electron energies by the electron–electron repulsion which has been counted twice and amounts to $17 \cdot 8$ eV.

Thus the binding energy of the molecule if both electrons occupy the symmetric bonding molecular orbital $\psi_a + \psi_b$ will be $-3 \cdot 7$ eV, compared with the experimental value of $-4 \cdot 7$ eV, and the calculation could in principle be improved by providing for spatial correlations between the electron positions.

Our experience with the operation of the Pauli principle in atoms makes it reasonable that both electrons can use ψ_+ if they have opposite spins. The total wavefunction Ψ will be antisymmetric if Ψ_σ the spin function is antisymmetric since

$$\Psi = \psi_+(r_1)\psi_+(r_2)$$
$$= \psi_a(r_1)\psi_b(r_2) + \psi_a(r_2)\psi_b(r_1) + \psi_a(r_1)\psi_a(r_2) + \psi_b(r_1)\psi_b(r_2) \quad (2.2)$$

and Ψ_σ must be $[(\uparrow_1, \downarrow_2) - (\downarrow_1, \uparrow_2)]/\sqrt{2}$ which is a spin singlet, just as (p. 13) in the Mg atom where two electrons occupy the same orbital, with opposed spins.

Note that the total spatial wavefunction (2.2) includes ionic terms $\psi_a(r_1)\psi_a(r_2)$ and $\psi_b(r_1)\psi_b(r_2)$ which put both electrons on one atom and this must clearly be wrong at large values of R_{ab}; there the total spatial wavefunction used by Heitler and London

$$\Psi = \psi_a(r_1)\psi_b(r_2) + \psi_a(r_2)\psi_b(r_1)$$

is more appropriate. (In their treatment the step beyond first-order perturbation theory, which yields

$$E = \int \Psi \mathcal{H} \Psi d\tau / \int \Psi \Psi d\tau$$

with $\Psi = \psi_a(r_1)\psi_b(r_2)$), was taken simply by allowing for exchange of the two electrons, i.e., by using $\Psi = \psi_a(r_1)\psi_b(r_2) \pm \psi_a(r_2)\psi_b(r_1)$.)

The type of bonding in H_2 described by the above treatment is ideal covalent bonding and it has a number of interesting features. It involves a pair of electrons of opposed spins; it saturates, in the sense that one cannot add a third H atom to the molecule without being forced to use high energy states to accommodate the extra electron; it cannot operate between the closed $1s^2$-shells in He since that would involve use of the energetically unfavourable antibonding states for two of the electrons (i.e. there is a closed-shell repulsion) and, finally, it leads to a characteristic energy versus nuclear separation relationship of the form shown in Fig. 2.3 which underlies discussions of vibrational excitations of molecules and which leads chemists to speak of a characteristic covalent bond radius†.

---

†For atoms with full inner shells the energy of a pair rises rather steeply when R_{ab} is reduced to the point where ion core overlap begins and this makes possible the definition of an ion core radius which is normally appreciably smaller than covalent or metallic radii.

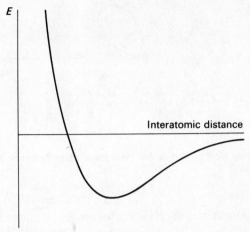

Figure 2.3   The binding energy of a diatomic molecule as a function of internuclear separation.

Covalent bonding of pure s-character is rare: although molecules of the type Li_2, Na_2, Cu_2 do exist, they are less stable than the crystalline metals which, as we shall see, are giant N-atomic molecules bonded by the more generalized electron sharing of the metallic bond. When the principal quantum number n is greater than 1 elements with ground state ns^2 configurations *can* form covalent bonds with molecular orbitals that make use of low-lying np atomic orbitals, and we must now consider the bonding behaviour of p-orbitals.

2.3   Bonding with p- and d-functions

The linear combination of atomic orbitals leads to bonding because the overlap of atomic wavefunctions leads to large $|\psi|^2$ in potentially attractive regions. When we use atomic orbitals that are not spherically symmetric the shapes of the wavefunctions impose a directionality on the bonding if overlap is to be maximized, and for wavefunctions like the $n = 2$, $l = 1$, functions of Fig. 1.4, which have odd parity, i.e. $\psi(-r) = -\psi(r)$, care will required to match the signs since for the molecular orbital the probability (and charge density) are proportional to $|\psi_a + \psi_b|^2$.

In molecules bonded by p-functions the potential in which electrons move cannot be regarded as one of spherical symmetry plus a small axial field, and the basic set of three p-functions that are used in discussions of bonding are (see p. 5) p_x, p_y, p_z functions which have the forms $xf(r)$, $yf(r)$ and $zf(r)$ where f is some radial function. These (like the $m_l = 0, \pm 1$ states which have the form $zf(r)$,

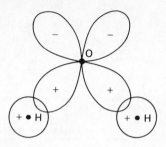

Figure 2.4   Overlapping atomic orbitals that yield the bonding molecular orbitals of H_2O. The sign of ψ is indicated.

$(x \pm iy)f(r))$ are eigenstates of the Hamiltonian for the atom, but not of the component of angular momentum m_l. They are the appropriate kinds of basis state and have the great advantage of providing an understanding of the shapes of many molecules. Thus in H_2O the p_x and p_y orbitals of oxygen overlap, as shown by the boundary surfaces in the Fig. 2.4, with 1s orbitals of two H atoms to produce two molecular orbitals whose charge clouds are approximately at right angles to one another. Since O has the electronic configuration $(1s^2)(2s^2)$ $2p^4$ only two empty states in the p-shell exist and one p-orbital (say p_z) must contain a pair ($\uparrow\downarrow$) of electrons that cannot take part in the bonding.

Where both atoms in a molecule have an unpaired p-electron two types of wavefunction overlap are possible as represented in Fig. 2.5. The first, where the axis of the charge cloud of the bonding orbital is that of the molecule is called a σ bond and the second is called a π bond (see Fig. 2.9). The bonding in the F_2 molecule is of the σ type and can be regarded as approximately

$$F_a \; 1s^2 \, 2s^2 \, 2p_x^2 \, 2p_y^2 \, \boxed{2p_z\uparrow \qquad 2p_z\downarrow} \, 2p_y^2 \, 2p_x^2 \, 1s^2 \, F_b$$

where the molecular axis is the z direction and the $2p_z$ wavefunctions overlap to give bonding (opposed spins) and antibonding (parallel spins) orbitals of which only the former is occupied; although a more correct description would recognize that bonding and antibonding orbitals of π character are formed by the p_x and p_y functions, and all four of these are occupied by electron pairs (Fig. 2.6); the separation of the energy levels in the centre of the figure is that for the molecule at its observed atomic spacing.

The splitting of the bonding and antibonding states is less for the π bonds than for σ bonds because of their smaller overlap; also the antibonding π_x and π_y orbitals constitute a degenerate pair (as do the bonding π_x and π_y orbitals). Thus from this viewpoint it is easy to see that if we remove one nuclear charge and one electron from each atom to produce the O_2 molecule, the two antibonding π_a orbitals have only two electrons between them. Just as in an

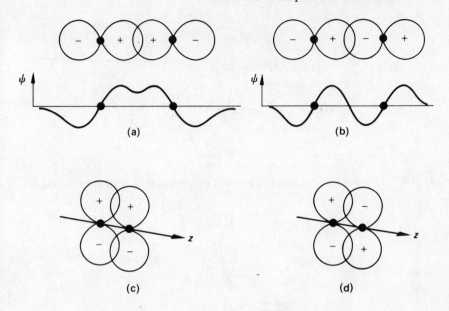

Figure 2.5   p–p overlap to give: (a) σ_b bonding orbital; (b) σ_a antibonding orbital;
(c) π_b bonding orbital; (d) π_a antibonding orbital.

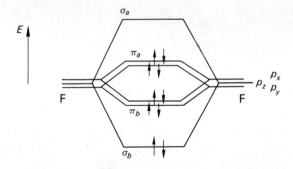

Figure 2.6   Energy levels for atomic orbitals in F atoms (left and right) and for molecular
orbitals in F_2 (centre): (a) antibonding; (b) bonding. Pairs of electrons occupying orbitals
are indicated by pairs of arrows. The three atomic levels are degenerate as are the pairs of π_a
or π_b levels.

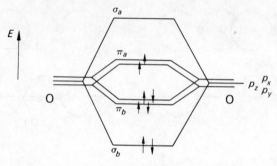

Figure 2.7   Formation and occupation of molecular orbitals in O_2. (See legend to Fig. 2.6.)

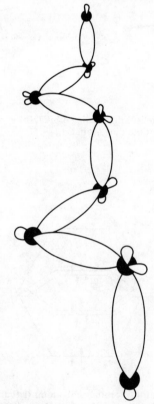

Figure 2.8   Long chain molecules of Se formed by p σ bonds (2 per atom). The charge clouds of only the bonding orbitals are shown.

atom (Section 1.4) the electron–electron repulsion is lessened if one electron enters each orbital, and the electrons have parallel spins (Fig. 2.7). The paramagnetism of O_2 is thus a molecular orbital analogue of the Hund's rule paramagnetism of atoms with part-filled shells†. The bonding in described as one σ bond plus two half π bonds; because double occupancy of a bonding π orbital is a whole bond and double occupancy of both bonding and antibonding π orbitals is no bond (as in F_2), double occupancy of the bonding orbital and single occupancy of the antibonding orbital can be called a half bond.

Because π bonds are effective only if the interatomic distance is small they are not found in heavier elements of group VI which have large ion cores. In the solid state both Se and Te form two σ bonds (Fig. 2.8) at approximately 90° to one another with consequent long chain molecules, although S can achieve a compromise closed molecule by closing its chain into an 8-membered ring, the S_8 molecule.

The charge cloud distributions of σ_b and π_b bonds are shown schematically in Fig. 2.9.

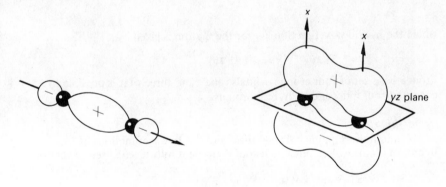

Figure 2.9   Details of σ and π bonds.

A very important type of covalent bond involves the construction of molecular orbitals which, for at least one of the bonded atoms, use a linear combination of orbitals of different l, this process being described as hybridization. Hybridization plays a vital part in the chemistry of C where the molecular orbitals bonding C to, say, H must often be written as (see p. 26)

$$\psi_{m.o.} = p\psi_h + q\psi_H$$

†Since, as we shall see later (p. 83), the 3d electrons of transition metals are best described by states related to LCAO molecular orbitals, the appearance of ferromagnetism at the end of the series where antibonding orbitals must be occupied is a closely related phenomenon.

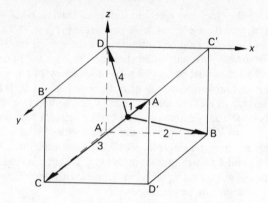

Figure 2.10   Axes of sp^3 hybrid orbitals. The arrows are labelled with the subscripts of the appropriate atomic hybrid orbital given in the text. The CH$_4$ molecule is formed by placing H atoms at A, B, C and D.

where the hybrid wavefunction ψ_h for the C atom is given by

$$\psi_h = a\psi_{2s} + b\psi_{2p_x} + c\psi_{2p_y} + d\psi_{2p_z}$$

From a basis set of four atomic orbitals (one s and three p) it is possible to construct four independent hybrid orbitals.

(i) sp^3-hybrids

Maximum overlapping between these and the 1s-wavefunctions of four hydrogen atoms will be possible if four equivalent hybrids are used of the form

$$\psi_1 = \psi_{2s} + \psi_{2p_x} + \psi_{2p_y} + \psi_{2p_z}$$

$$\psi_2 = \psi_{2s} + \psi_{2p_x} - \psi_{2p_y} - \psi_{2p_z}$$

$$\psi_3 = \psi_{2s} - \psi_{2p_x} + \psi_{2p_y} - \psi_{2p_z}$$

$$\psi_4 = \psi_{2s} - \psi_{2p_x} - \psi_{2p_y} + \psi_{2p_z}$$

since (see Fig. 2.10) these give charge clouds pointing towards the corners of a regular tetrahedron.

Notice that the ground state of the C atom is s^2 p^2 and the sp^3 hybrid molecular orbitals require the participation of the excited sp^3 atomic configuration, but since this in the ^{5}S$_0$-state has considerable Hund's rule type stabilization, the cost in energy is easily compensated for by the extra bonding energy provided in the formation of methane CH$_4$.

(ii) sp²-hybrids

It is possible to construct molecular orbitals that are co-planar by hybridization between the 2s-orbital and two (say p_z and p_y) of the p-orbitals of C. A symmetrical charge distribution in the *zy* plane is given by the three trigonal orbitals

$$\psi_1 = \frac{1}{\sqrt{3}}(\psi_{2s} + \psi_{p_z}\sqrt{2})$$

$$\psi_2 = \frac{1}{\sqrt{6}}(\psi_{2s}\sqrt{2} + \psi_{p_y}\sqrt{3} - \psi_{p_z})$$

$$\psi_3 = \frac{1}{\sqrt{6}}(\psi_{2s}\sqrt{2} - \psi_{p_y}\sqrt{3} - \psi_{p_z})$$

which have the boundary surfaces shown schematically in Fig. 2.11. The odd parity of the p-functions makes these orbitals asymmetric with respect to the nucleus.

Two C atoms 'prepared' in the configurations $p_x^1(sp^2)^3$ can come together as in Fig. 2.11 with four H atoms to form the ethylene molecule, with a (sp^2) hybrid bond of σ type between the two carbon atoms. Since, however, the singly occupied p_x-orbital of each C atom will be normal to the plane of the molecule a π bond can also be constructed so that we may speak of the C atoms as having a double bond between them.

(iii) Ring bonds

A very important molecule in the history of theoretical chemistry is benzene, C_6H_6, which is known to form a planar ring. That is clearly possible with trigonal sp^2-hybrids arranged as shown in Fig. 2.12a, and the long discussed problem of how to arrange the extra bonding electrons is solved when one recognizes that the p_x-orbitals normal to the plane of the ring can be combined on LCAO principles to form three bonding and three antibonding orbitals. One of the bonding orbitals is

$$\psi = \psi_{p_x}^1 + \psi_{p_x}^2 + \psi_{p_x}^3 + \psi_{p_x}^4 + \psi_{p_x}^5 + \psi_{p_x}^6$$

(where the superscript numbers label the atoms) which gives doughnut-shaped charge clouds above and below (Fig. 2.12b) the ring (with ψ positive above, negative below); the simplest antibonding orbital is

$$\psi = \psi_{p_x}^1 - \psi_{p_x}^2 + \psi_{p_x}^3 - \psi_{p_x}^4 + \psi_{p_x}^5 - \psi_{p_x}^6$$

which, like the antibonding molecular orbital of H_2, gives a zero for ψ between adjacent pairs of atoms. The other four orbitals are slightly more complicated linear combinations of the atomic p_x orbitals.

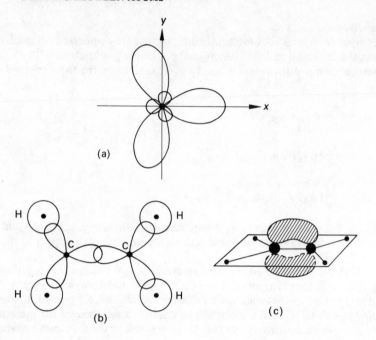

(a)

(b)   (c)

Figure 2.11   (a) Boundary surfaces in the xy plane for sp^2 hybrid orbitals. (b) The ethylene molecule C_2H_4. (c) The C–C π bond in C_2H_4 (the sp^2 σ hybrid molecular orbitals are shown only by their axes).

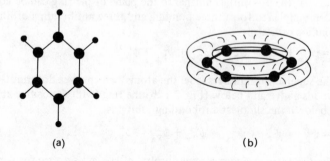

(a)   (b)

Figure 2.12   (a) sp^2 hybrid orbitals in C_6H_6 (bond axes only). (b) Ring (π) orbital of lowest energy in C_6H_6.

These ring orbitals are conceptually very significant as a bridge between the two-centre orbitals of simple covalent bonds and the N-centre orbitals we shall invoke in solids. The simplest example of a ring orbital however is $(\psi_a + \psi_b + \psi_c)$ ($\psi(1s)$ for each atom) on an equilateral triangle of H atoms; occupation of it by two electrons yields the stable ionic species H_3^+ which needs $4 \cdot 5$ eV to break up into $H_2 + H^+$.

(iv) d-bonding

Pure covalent d-bonds are very uncommon, largely because the one-electron energy of an nd-orbital in a transition metal atom is close to that of $(n + 1)$ s- and $(n + 1)$ p-levels. Hybrid dsp-bonding is therefore expected, and found, and some very impressive predictions of the shapes of transition element compounds have been made by Pauling and his followers. We shall not discuss any of these, but it should be borne in mind that such hybridization effects must be expected to play a role in metals and alloys of the transition series.

2.4   The metallic bond

Because of the simple crystal structures of metals it seems likely that the component atoms are held together by a very general sharing of their outer electrons, and we shall be devoting Chapters 3 and 4 to a discussion of the characteristics of the states occupied by these electrons. We shall see that the source of the binding energy in metals and covalently bonded materials is essentially the same, the overlap of atomic wavefunctions to produce molecular or crystal orbitals in which electrons are accommodated at lower energy than in the separated atoms. The striking difference in the response to applied electric fields shown in the vastly different conductivities of diamond and copper will be seen in Chapter 4 to follow from the distribution in energy of allowed and occupied orbitals. The only distinction that can normally be made about the *spatial* distribution of the bonding electrons is that the charge cloud density we associate with metallic bonding is rather uniformly distributed through the crystal lattice, whereas in diamond a resemblance to the (sp^3) hybrid covalent bonds of CH_4 is maintained by a concentration of the charge density along the lines joining nearest-neighbour atoms in the tetrahedral structure. However, the three-dimensional network of such covalent bonds in a C (diamond) crystal, will mean that the crystal (molecular) orbitals extend over every atom in this giant molecule.

The difficulty of maintaining a sharp distinction between covalent and metallic bonding is also brought out by considering the graphite form of carbon (Fig. 2.13). In the two-dimensional sheets the ring (π) orbitals of benzene have linked to extend throughout the sheet, just as the (sp^2) planar orbitals do. Weaker bonding by interaction of the ring orbitals associated with neighbouring sheets, and little directional character for this bonding, are indicated by the ease

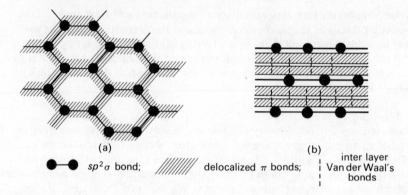

Figure 2.13   Bonding in graphite showing: (a) the two-dimensional sp² bond network; (b) the delocalized charge clouds of the π bonding electrons (section perpendicular to the sp² sheets).

with which these sheets slide over each other to yield the familiar lubricative properties. It seems natural to regard the bonding contribution of the ring orbitals as more metallic than covalent (but it is by no means obvious that an overlap of bonding and antibonding orbitals will take place to yield metallic conductivity). However, a smaller separation between bonding and antibonding π crystal orbitals than between bonding and antibonding σ orbitals (sp³ or sp²) would be in accord with our comment on such a difference for π and σ molecular orbitals in F_2 (p. 33, Fig. 2.6). A schematic indication of the energy levels is given in Fig. 2.14.

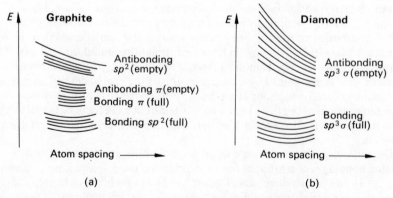

Figure 2.14   Schematic representation of crystal molecular orbital energy levels as a function of atomic spacing for values near the observed spacing: (a) graphite (sp² σ bonds and delocalized π bonds); (b) diamond (tetrahedral network of sp³σ bonds only).

Just to emphasize that it is the energies of the bonding and antibonding orbitals, rather than the spatial charge distribution, that are important in determining the electrical properties of a crystal, consider the example of Ge. Although crystalline Ge has the same structure as diamond, with tetrahedral coordination, the valence electron charge distribution (both as calculated and as measured experimentally) is spatially almost uniform throughout the unit cell, and the electrons cannot be 'seen' to be localized into bonds. Pure Ge is an insulator (at low temperature) because these electrons are in fully occupied bonding orbitals that are separated by a finite gap from empty antibonding orbitals. Uniformity of charge distribution must not be taken to imply metallic conducting behaviour.

It is worth remarking at this point that two molecular orbitals were yielded for *each* atomic orbital of the two atoms in H_2. In crystals (covalent or metallic) we might therefore expect nN orbitals, half of them effectively bonding and half antibonding where rather general overlap of n atomic orbitals has taken place in a crystal of N atoms.

2.5 Other contributions to bonding in solids

2.5.1 Van der Waals' bonding

We have referred to the saturation of covalent bonding such that H_2, He, Ar, O_2, etc., should be incapable of bonding to other atoms. But we know that the molecules of these gases will bond into crystals at *low* temperatures. Since the intermolecular interactions involved are the same as those making necessary corrections to the perfect gas laws, this bonding is called van der Waals' bonding. If the molecules possessed electric dipole moments polarization forces would be expectable, but in symmetric molecules it is necessary to invoke a synchronization of fluctuations into polarized states in adjacent molecules as the source of van der Waals' bonding. In quantum mechanical terms one requires the admixture into the molecular orbitals describing the charge clouds of the outer electrons of excited state wavefunctions which are less symmetric. Perturbation theory treatments† yield reasonable orders of magnitude (0·01 eV for Kr at 4 Å separation) and a $1/r^6$ dependence on distance. The more extensive the electron cloud, the more polarizable the molecule and so we expect stronger van de Waals' bonding in heavier atoms. Thus the heat of sublimation (into molecules) of solid iodine is 0·8 eV as compared with about 0·2 eV for chlorine crystals. Because of the non-directional nature of the interaction we expect close-packed structures for spherically symmetric molecules.

† For details see Schiff, *Quantum Mechanics* (McGraw-Hill) 1955.

2.5.2   Ionic and 'mixed' bonding

We have already referred briefly to ionic crystals; for the simplest examples the bonding can be discussed in terms of classical Coulomb forces without reference to the wavefunctions of the occupied or empty states. Consideration, however, of our approach to molecular bonding shows that for heteronuclear molecules the molecular orbitals will have wavefunctions of the form $\psi_m = \psi_a + \lambda\psi_b$, and when two electrons provided by atoms a and b occupy this orbital the bonding will change from ideally covalent to ideally ionic as λ goes from one to zero. Correspondingly, in the sequence of solids

$$Ge \rightarrow GaAs \rightarrow ZnSe \rightarrow CuBr$$

increasing ionic character is clearly manifesting itself, and the preservation of the tetrahedral coordination of A atoms by B atoms should not lead   one to invoke transfer, say, of two electrons from $s^2 p^4$ Se to s^2 Zn to make possible the construction of traditional sp^3-hybrids. All discussions of solid state bonding in terms of specific percentages of ionic, covalent, metallic and van der Waals' bonding should be treated with great caution; in the hands of intuitive chemists they can be illuminating, they can also be misleading. In all systems however, as the atomic number increases in a given column of the periodic table the electron clouds become larger relative to interatomic spacings and overlap effects play an increasing role. We shall see some of the consequences in the structures and bond energies of the elemental solids in § 2.6. In compounds of heavy elements 'pure' ionic and 'pure' van der Waals' character are never found and the heaviest 'inert' gas Xe has recently been found to be by no means chemically inert.

2.5.3   Hydrogen bonding

Although it plays no role in most of the materials traditionally considered as subject to the rule of solid state physics, there is one other type of bonding that although weak ($\sim\frac{1}{4}$ eV) is of great importance in some substances, including water and many organic molecules such as proteins and DNA. It does not involve sharing or transfer of electrons but rather sharing of protons. It is represented in Fig. 2.15 by dotted lines where normal covalent bonds are shown as full lines. Detailed discussions of hydrogen bonding can be found in text-books of structural and theoretical chemistry. The essential feature of the bond is the presence of a small electropositive atom on the line joining two fairly electronegative atoms so as to maximize the electrostatic coupling between the molecules. As electropositive atom only H seems small enough to allow an approach close enough to yield significant bonding, and as electronegative atom O, N, F and occasionally S are found to be involved. These effects are not easy to calculate rigorously and it is likely that contributions are also made by covalent effects – that is, by admixture into the total wavefunction of the system of a small amount of a molecular orbital distributed along the dotted lines of Fig. 2.15.

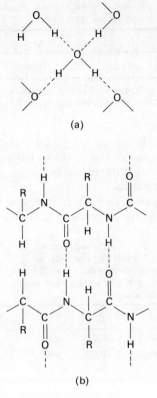

(a)

(b)

Figure 2.15   Hydrogen bonding: (a) in ice (note the open structure); (b) between long chain molecules (R is a group of the form C_nH_{2n+1}).

2.6   Bonding energies and structures in the solid elements

The structures of the elements have been presented in terms of concepts like those outlined above in the volume in this series by Brown and Forsyth, but it may strengthen the generalizations we have made if we compare for some of the groups and periods of the periodic table the magnitudes of the energies involved. We shall express our energies throughout in electron volts, but the traditional chemical literature used kilocalories per gramme atom, numbers in the latter system needing to be divided by 23·1 to yield electron volts/atom. Table 2.1 gives the heats of sublimation of the solid close-packed rare gases, and although the value for He is anomalously low for special reasons, the steady increase with atomic number is striking, as is that of the boiling point. H, O and N all solidify as molecular crystals with the identity of the molecular unit being preserved and

Table 2.1   Experimental cohesive energies (heats of sublimation) for the crystalline inert gases.

Element	He	Ne	Ar	Kr	Xe
ΔH_s(eV/atom)	0·002†	0·02	0·08	0·11	0·16

†Helium does not become solid at any temperature under normal pressures; this value is the heat of evaporation from the liquid.

we can compare (Table 2.2) the energy ΔH_s to separate the crystal into a gas of molecules and D the heat of dissociation of the molecules. The ratio d_m/d_a of the separation of the molecules in the crystal and of the atoms in a molecule is also indicative of a clear distinction.

When, however, we compare (Table 2.3) the halogens the sharp distinction between the binding of atoms into molecules and of molecules into crystals is

Table 2.2   Heats of sublimation (crystal → molecule) and heats of dissociation (molecule → atoms) for the diatomic gases, indicating weak van der Waals' bonding.

Element	H_2	N_2	O_2
ΔH_s(eV/atom)	0·004	0·035	0·04
D(eV/atom)	2·25	4·9	2·6
d_m/d_a	5	3	3

no longer present. Furthermore, I is a solid at room temperature (where F and Cl are gases), it possesses a strong colour with a distinct quasi-metallic lustre and is found to become a metallic conductor under hydrostatic pressures accessible in the laboratory. Clearly, states of the outer electrons which extend from molecule to molecule through the solid are available at fairly modest excitation energies. Notice not only that ΔH_s increases from Cl to I but also that D decreases. Very crudely, one can say that as the charge clouds associated with

Table 2.3   The same as Table 2.2 for the halogens, indicating stronger van der Waals' bonding.

Element	F_2	Cl_2	Br_2	I_2
ΔH_s(eV/atom)	0·04	0·13	0·22	0·39
D(eV/atom)	0·82	1·25	1·0	0·78
d_m/d_a	2·1	1·5	1·45	1·3

Table 2.4.   Bonding energies in the group VI (chalcogenide) elements. The heat of dissociation of diatomic molecules D_{x_2} is given as a measure of the strength of covalent bonding and the difference between this and the energy ΔH_{at} required to dissociate the crystal into individual atoms is given as a measure of the strength of the van der Waals' bonding between quasi-molecular units.

Element	O	S	Se	Te
$\Delta H_{at} - D_{x_2}$(eV/atom)	0·04	0·7	0·7	0·85
D_{x_2}(eV/atom)	2·6	2·2	1·4	1·1
d_m/d_a	3·0	1·6	1·5	1·2

electrons involved in van der Waals' bonding or covalent bonding become less localized (more 'metallic') in character the former bond increases in strength and the latter decreases. The strength of the solid depends on its weakest link (the van der Waals' bonding) so the boiling point increases with atomic number.

A similar situation holds for the series O—S—Se—Te (Table 2.4) but, as we saw on p. 35, the structures change from closed molecules, and the quantities we have plotted are $\Delta H_{at} - \Delta D_{x_2}$ (the difference in the energies required to atomize the solid and the dissociation energy of a diatomic molecule) as a measure of the binding together of the molecules, and ΔD_{x_2} as a measure of the 'covalent' bonding. The same quantities are given for P, As, Sb and Bi in Table 2.5.

In contrast to all the other groups, C, Si, Ge and grey Sn show a heat of sublimation (from crystal to atoms) (Table 2.6) that decreases with atomic number. This is understandable when we note that their three-dimensional tetrahedral networks do not require the use of van der Waals' bonds. In groups V, VI and VII the latter constitute the weak links in the crystal bonding, and since as we have seen they become stronger as the size of the molecular charge clouds increase the heat of sublimation increases with atomic number; in group IV increasing size of the bonding charge cloud (corresponding to decreasing

Table 2.5   The same as Table 2.4 for Group V (pnictide) elements.

Element	P	As	Sb	Bi
$\Delta H_{at} - D_{x_2}$(eV/atom)	0·9	1·0	1·1	1·1
D_{x_2}(eV/atom)	2·5	1·9	1·6	0·96
d_m/d_a	1·8	1·25	1·17	1·12

Table 2.6   Heats of sublimation (here a measure of the covalent/metallic bond strength) for elements in group IV.

Element	C	Si	Ge	Sn
ΔH_s(eV)	7·35	3·7	3·7	3·4

directionality of   the bonds and in our terminology increasing metallic character) is associated with weakening of the bonding.

For simple metallic systems fairly straightforward trends in cohesive energies with number of electrons involved in bonding are evident (Table 2.7), but for transition metals it is not clear what this number is. The reason is apparent from Fig. 4.23 where we give schematically the modification of the outer electron energy levels as the atoms are brought together into the crystal. Table 2.7 shows that the d-electrons make important contributions to bonding

Table 2.7   Heats of sublimation for some of the metallic elements. Note that where n_v, the number of bonding electrons, is unambiguous the range of variation in $\Delta H_s/n_v$ is fairly small (½ – 1½ eV/atom per electron).

Element	K	Ca	Sc	Ti	V	Cr	Mn	Fe	Co	Ni	Cu	Zn
ΔH_s(eV/atom)	0·93	1·8	3·9	4·9	5·3	4·1	3·0	4·1	4·4	4·4	3·5	1·3

Element	Rb	Sr	Y	Zr	Nb	Mo	Tc	Ru	Rh	Pd	Ag	Cd
ΔH_s(eV/atom)	0·84	1·7	4·4	6·3	7·5	6·8	6·0	6·7	5·8	3·9	2·9	1·0

Element	Cs	Ba	La	Hf	Ta	W	Re	Os	Ir	Pd	Au	Hg
ΔH_s(eV/atom)	0·82	1·8	4·5	6·5	8·1	8·6	8·1	8·1	6·9	5·8	3·8	0·65

but that this falls as the d-shell becomes more than half full. The chemist's language of hybridized dsp-bonding seems not wholly inappropriate for much of the series but fails badly for Cu, Ag and Au, where physical evidence (see Chapter 5) makes it overwhelmingly clear that d-levels are full, but the cohesive energy suggests that they still play a part in bonding, being larger than in Ca, Sr and Ba on the one hand and in Zn, Cd and Hg on the other. (The particularly small value for Hg is interpreted by the chemist as indicating a tendency towards a closed $6s^2$ subshell – the inert pair – in the third long period metals Hg → Bi.)

3

Electrons in Metals:
The Free-Electron Gas

3.1 Introduction

In our discussions so far of electron states in molecules and solids we have considered states which are essentially produced by the overlap of atomic wavefunctions and our nomenclature has preserved some degree of reference to the character of the atomic states. We shall return to this approach later even for metallic systems (especially those containing transition elements), but historically the first detailed discussion of electron states in solids was for metallic solids, and was in terms of electron states which were regarded as eigenstates of the whole crystal. Such an approach does not easily yield either a physical picture or numerical values for bonding energies of metals, but is an invaluable starting point for discussions of conductivity and related characteristically metallic properties and it will be the basis of the next two chapters. (Notice however that in discussions of the energy eigenstates of 'metallic' electrons it is possible in principle to express the energy relative either to the lowest energy state of an electron outside the metal or to the energy states of electrons in ion core states of the atoms in the crystal. In solid state spectroscopy one uses energy scales of the latter sort and questions of the symmetry character of initial and final states will enter into transition probabilities.)

3.2 Eigenstates of a free-electron gas

The simplest model, then, of a metal takes the crystal as a structureless flat-bottomed potential well (Fig. 3.1) the linear dimensions of which are those of the crystal and the depth (V) of which is great enough to ensure that the wavefunctions of all states in which we will be interested will fall off very rapidly outside the box. It must be noted that the wavefunctions we are concerned with are one-electron wavefunctions (as in the self-consistent field treatment of the many-electron atom) and the potential we are treating as constant is that due to all the ion cores and all the other conduction electrons.

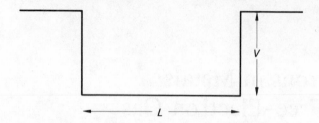

Figure 3.1   The potential well of the free-electron model.

The eigenstates of such a model in one dimension are easily obtained from the Schrödinger equation

$$\nabla^2 \psi - \frac{2m}{\hbar^2} (E - V(x)) \psi = 0 \tag{3.1}$$

(The problem is identical to that of the electron-in-a-box model for a H atom or a nucleon in a nucleus except that our box may be 1 cm in length rather than 10^{-8} or 10^{-13} cm! The energy level spacing is correspondingly drastically different.)

In principle we can use the standing wave solutions of the form $\psi = \sin kx$ which are allowed (see Fig. 3.2) for box-like boundary conditions ($\psi = 0$ at $x = 0$ or $x = L$), and these would yield k values of $(2\pi/L) n/2$ and allowed energy levels $E_n = (\hbar^2/2m) k_n^2$. In three dimensions the wavevector k replaces k as the quantum number (with three spatial components) that specifies an allowed state. If L is of macroscopic dimensions it is clear that, with about one electron per atom to be accommodated, the number of such one-electron eigenstates occupied (with two electrons of opposed spin) will be very large, and the spacing in energy of successive levels extremely small.

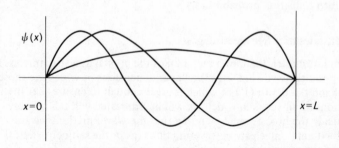

Figure 3.2   Sine wave solutions to Schrödinger's equation for an infinitely deep potential well.

In many situations in metals, however, it is very much more convenient to use as our set of allowed states the plane wave states $\psi = \exp(ikx)$ (for one dimension) and $\psi = \exp(i\mathbf{k} . \mathbf{r})$ (for three dimensions) which are allowed for the cyclic boundary conditions

$$\psi(x) = \psi(x + L)$$

These boundary conditions are used also for the discussion of lattice vibrational waves in crystals (see W. Cochran, *The Dynamics of Atoms in Crystals*, number 3 in this series). The allowed values of k_x are now given by $k_x = (2\pi/L)n$ with n an integer, but the relationship between E and k is unchanged. This might seem to imply a reduction by a factor of two in the number of states in a given range of energies, but in fact k may now take both positive and negative values, so that the distribution of allowed states is exactly the same for our plane wave states as for standing wave states. Notice that the x-component of momentum p_x is given in quantum mechanics by

$$p_x \psi = \frac{\hbar}{i} \frac{\partial \psi}{\partial x}$$

and therefore the state $\psi = \exp(ik_x x)$ has momentum $\hbar k_x$ which is just the classical value mv_x if $\lambda = h/mv$ and $k = 2\pi/\lambda$. This is of course what we should have expected for an electron propagating freely in a constant potential. The standing wave state $\psi = \sin(k_x x)$ has no sharp value for p_x.

We can now use these free-electron states to calculate the maximum energy and other properties of N electrons occupying at $T = 0$ such states in a crystal cube of volume L^3. We will equate N to $n_v N_a$ where N_a is the number of atoms and n_v is the conventional metallic valency of the atoms. In a vector space (k-space) with axes k_x, k_y and k_z an allowed state is represented by a point k; associated with each state is a box of side $2\pi/L$. The lowest energy configuration of the system will accommodate the electrons in a sphere in k-space of radius k_{max} where $2(\frac{4}{3}\pi k_{max}^3) = N(2\pi/L)^3$ since each box has volume $(2\pi/L)^3$ and holds two electrons. Thus the maximum energy E_{max} of occupied states is given by

$$E_{max} = \frac{\hbar^2}{2m} k_{max}^2 = \frac{\hbar^2 \pi^2}{2m} \left(\frac{3}{\pi}\right)^{2/3} \left(\frac{N}{L^3}\right)^{2/3} \qquad (3.2)$$

that is to say, a function only of the number of electrons per unit volume N/L^3. We can express this in a convenient way as

$$E_{max} = 36 \cdot 1 (n_v/\Omega_a)^{2/3} \qquad (3.3)$$

where E_{max} is in electron volts, Ω_a is L^3/N_a the atomic volume expressed in cubic angstroms (10^{-30} m^3) and n_v the number of electrons per *atom*. We may rearrange Equ. 3.2 to give the number of electrons per unit volume

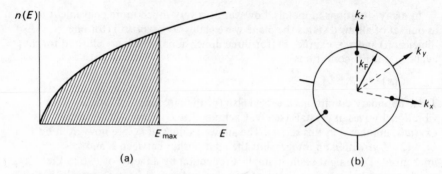

Figure 3.3 (a) The density of allowed electron states $n(E)$ for the free-electron model. (b) The spherical constant energy surface enclosing all occupied states.

accommodated in states up to a given energy E as

$$\frac{N}{L^3} = \frac{\pi}{3}\left(\frac{2m}{\hbar^2\pi^2}\right)^{3/2} E^{3/2}$$

so that the number of *states* per unit volume of crystal with energy less than E is

$$N_s(E) = \frac{\pi}{6}\left(\frac{2m}{\hbar^2\pi^2}\right)^{3/2} E^{3/2} \tag{3.4}$$

which may be differentiated with respect to the energy to yield the density of states $n(E)$ as a function of energy for the whole *band* of allowed states

$$n(E) = \left(\frac{1}{4\pi^2}\right)\left(\frac{2m}{\hbar^2}\right)^{3/2} E^{1/2} \tag{3.5}$$

The results of this simple model can therefore be summarized in Fig. 3.3 where shading indicates states occupied by N electrons.

The surface of the sphere in Fig. 3.3b is the surface of constant energy enclosing all occupied states, and is called the *Fermi surface*. Its radius is what we have called k_{max}, but we shall also use for this quantity the expression *Fermi wavevector* denoting it k_F.

Since the number of states per unit volume in a surface element dS of the spherical shell of thickness δk at the Fermi surface is $(1/8\pi^3)\,dS\delta k$ then

$$n(E)_{E_{max}} = \frac{1}{8\pi^3} \int\limits_{\substack{\text{Fermi} \\ \text{surface}}} \frac{dS}{dE/dk} \tag{3.6}$$

since $n(E)\delta E$ is the number of states per unit volume in an *energy* increment δE.

In real metals a constant energy surface $E = E_{\max} \equiv E_F$ enclosing all states occupied at $T = 0$ always exists, but is not in general spherical. Many such Fermi surfaces have now been established experimentally. Various properties of metallic solids depend on the quantities $E_{\max}$ and $n(E)_{E_{\max}}$ and we shall use experimental values of them as a test of the usefulness of this free-electron model.

3.3   Soft X-ray emission spectra and $E_{\max}$

The most direct experimental evidence for the broadening of sharp atomic levels into bands of levels in the solid state is provided by the observation of the broadening of certain types of spectroscopic emission lines in the solid state. The most useful energy range for such studies is in the soft X-ray region of the spectrum and a simple example is provided by the $3s \rightarrow 2p$ transition in the spectrum of Na. When, in the solid state, an electron has been removed from one of the 2p-levels of the $1s^2 2s^2 2p^6$ core of a particular Na atom, an X-ray quantum will be produced when an electron from the electron-gas band of levels (these are derived from the atomic 3s-levels) falls into the 'hole'. The X-radiation so produced will obviously not be monochromatic since the occupied free-electron levels are distributed over the range $0-E_{\max}$ (the bottom of this range is our arbitrarily chosen zero of energy) and this range will appear as the width of the emission band (see Fig. 3.4). A detailed account of the process whereby this

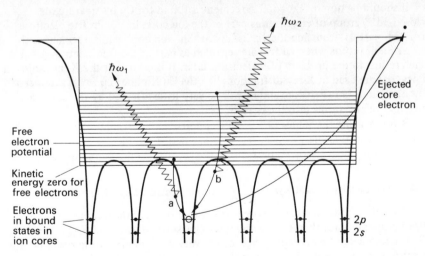

Figure 3.4   Soft X-ray emission in Na. Occupied conduction band levels are indicated by horizontal lines. If electron a comes from the bottom state in the band and electron b from the highest occupied state $E_{\max} = E_b - E_a$.

emission takes place will be rather complicated since: (i) an empty core level will obviously represent a local departure from the uniform potential we have assumed and will consequently distort the band in its vicinity; (ii) the emission intensity will vary not only with the density of occupied states but also with transition probabilities which depend on the detailed symmetry of the initial and final states of the electron; (iii) the electrons making such transitions may, in the initial electron-gas level, be involved in the collective excitations of the whole system.

Ignoring such complications however, we give the widths of such emission bands as the 'experimental' values of E_{max} in Table 3.1; the calculated values are given by Equ. 3.3. The value of n_v is the number of electrons outside the closed-shell ion core.

Table 3.1   Conduction band widths from the free-electron model and from X-ray emission spectroscopy (eV).

n_v Element	Li	Na	1 K	Cu	Au	2 Be	Mg	Zn	3 Al	4 Si	Ge	5(?) V
E_{max}(calc)	4·7	3·2	2·6	7·1	5·5	14·3	7·2	9·5	12·8	13	12	18
E_{max}(exp)	3·9	2·8	1·9	6·5	5·4	13·8	7·6	11·5	11·8	18	16	7

It should be noted that good agreement is found for Cu by treating the closed $3d^{10}$ group of electrons as part of the ion core; in V with five electrons — part 3d, part 4s — outside the $1s^2 2s^2 2p^6 3s^2 3p^6$ ion core, bad agreement with experiment results when all five are regarded as part of the free-electron gas. We shall return to the problem of d-electrons later. It is clear that the X-ray emission lines that are used in X-ray diffraction, like the Cu $K\alpha$ line (2p → 1s) produced by transitions between core states, will be very little affected by the coming together of the atoms into the crystal.

Before leaving the values of E_{max} it is interesting to estimate the velocity of electrons with this energy on the assumption that it can be equated to the kinetic energy of the most energetic particles in the gas; for Cu this is about 10^6 m s^{-1}.

3.4   The density of states $n(E)_{E_{max}}$ and the electronic specific heat

The idea of free-electron energy levels whose occupation is subject to the Pauli principle leads then to experimentally reasonable values of E_{max}. It also leads immediately to an explanation of the very small contribution made by the electrons to the specific heat of a metal, a contribution that could not be understood in terms of classical free-electron gas models developed to discuss conductivity. Thus we can see from Fig. 3.3a that only a small number of electrons, of the order of $n(E)_{E_{max}} k_B T$, in a narrow range near E_{max} have available to them empty states to which they can be excited by thermal energies of the order $k_B T$. We can therefore understand that the specific heat will be changed from the classical value by a factor of about T/T_F where $k_B T_F = E_{max}$. We shall see later that $n(E)_{E_{max}}$ plays an important role in all low energy properties of metals, and it is therefore useful that we can derive good estimates of it from the specific heat.

The formal statement that embodies the ideas of the preceding paragraph is that electrons obey Fermi–Dirac statistics in consequence of their spin and the Pauli principle. In the derivations of Section 3.2 we assumed that the system was at 0 K. We are now acknowledging that at any finite temperature the probability of occupation of a state of energy E is given by the Fermi–Dirac distribution function

$$f(E) = 1/[\exp(E - \mu)/k_B T + 1] \tag{3.7}$$

where μ, the thermodynamic potential which is uniform throughout a system in equilibrium, is given by the condition

$$2 \int_0^\infty n(E) f(E) dE = N \tag{3.8}$$

where N is the number of particles in the system (the factor 2 comes from the spin degeneracy of the electrons). Since at 0 K all states are occupied below E_{max} and empty above it is obvious that

$$\mu_0 = E_{max}^0 \dagger$$

for our free-electron gas. This condition does *not* hold for solids other than metallic conductors at 0 K and it must not be used in place of Equations 3.7 and 3.8 as a definition of μ.

In Fig. 3.5 we show the form of the Fermi–Dirac function for various temperatures, and in Fig. 3.6 the occupied states at a finite temperature. The latter shows that E_{max} is not well defined except at $T = 0$, but also suggests a simple approximation to the calculation of the electronic specific heat. By raising the temperature we have transferred electrons from the area SQP to the area PRT, and if we approximate the curve SPT with the straight line S'PT' we have a number of electrons $2 \times \frac{1}{2} PR \times RT'$ whose energy has been increased, on average, by $\frac{2}{3} RT'$. Now a straight line approximation to the tail of $f(E)$ gives RT' as about $2 \cdot 5 k_B T$, so that the total increase in energy U is about

$$2 \times (n(E)_\mu/2) \times \frac{1}{2} \times 2 \cdot 5 k_B T \times \frac{2}{3} \times 2 \cdot 5 k_B T = (25/12) k_B^2 T^2 n(E)_\mu$$

$\dagger \mu$ is sometimes referred to as the Fermi level, and E_{max}^0/k_B is called T_F, the Fermi temperature.

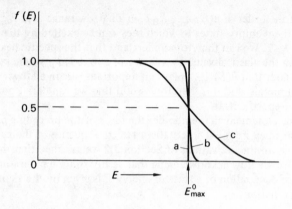

Figure 3.5   The Fermi–Dirac function for: (a) $T = 0$; (b) $E^0_{max}/k_BT \sim 40$; (c) $E^0_{max}/k_BT \sim 6$. For Ag (b) holds at the melting point.

where we write $n(E)_\mu$ for $n(E)$ at E equal to μ, giving

$$c_v = \frac{dU}{dT} \simeq 4\,k_B^2 n(E)_\mu T$$

(An exact derivation gives $c_v = (\pi^2/3)\,k_B^2 n(E)_\mu T$.)

It should be emphasized that this relationship will hold for metallic systems even when $n(E)$ is not given by the simple parabolic relationship of Equ. 3.5, provided that $n(E)$ does not vary strongly with E over a range k_BT. For this reason it is always best to express specific heat results in terms of $n(E)_{E_{max}}$

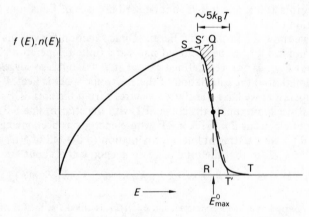

Figure 3.6   The density of occupied states $f(E)n(E)$.

Table 3.2   Experimental and theoretical (free-electron) values
of the electronic specific heat coefficient γ ; units are
mJ mol^{-1} K^{-2}

	Na	Al	Cu	Co	Ge	Si
γ_{exp}	1·4	1·35	0·70	4·73	0	0
γ_{fe}	1·1	0·91	0·50	0·35	1·0	0·9

rather than in terms of the free-electron model with modified values of its
parameters.

The electronic specific heat (γT) can be separated at low temperatures from
the lattice specific heat (which is proportional to $(T/\theta_D)^3$) by plotting c_v/T
which is given by

$$c_v/T = \gamma + \beta T^2$$

Some experimental values of γ are compared in Table 3.2 with those given by
Equations 3.3, 3.5 and the relation between density of states and electronic heat
capacity. For Si and Ge things have clearly gone wrong and for transition metals
the experimental values are very large.

By applying a magnetic field H rather than a temperature T to an electron gas,
one perturbs the energies of the levels by amounts $\mu_B H$ where μ_B is the Bohr
magneton. A simple argument (see J. Crangle, *Magnetic Properties of Solids*,
number 6 in this series) yields a magnetic susceptibility

$$\chi = 2\mu_B^2 n(E)_{E\,\text{max}}$$

and this small and temperature-independent (Pauli) susceptibility is in fact
observed in simple metals. It is often, however, enhanced by effects which are
less important in the specific heat, effects which can, in some limit, lead to a
spontaneous magnetization of the electron gas, *i.e.* ferromagnetism.

3.5   Electrical conductivity

One of the main aims of the early electron gas theories of metals was to
provide a satisfactory account of conductivity. This topic is dealt with at length
in another book in this series (that by J. S. Dugdale) but this chapter would be
incomplete without an indication of how the model we have outlined for the
electronic structure is used as a basis of such discussions.

The equilibrium distribution of electrons in a metal carries no current, for to
do otherwise would violate symmetry. When a steady electric field is applied the
occupation of states does change and a current is produced. The response of a
band electron to an electric field is not the same as that of a classical particle,
except in the special case of a band describing free electrons.

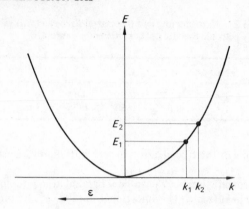

Figure 3.7   The effect of an electron field on a band electron; the initial state k_1 is accelerated to k_2 after time δt. The acceleration is in the opposite direction to $\mathscr{E}$ because of the negative charge on the electron.

We have seen that a definite momentum $\hbar k$ is associated with our state $\psi = \exp(ik \cdot r)$; if we wish to discuss electron transport and scattering we need also to consider the individual electron as a somewhat localized wave packet (whose dimensions are large on an atomic scale, but small on a macroscopic scale), its transport velocity being its group velocity v_g.

In Fig. 3.7 consider an occupied state k_1 having energy E_1; the group velocity v_g is $(1/\hbar)\mathrm{d}E/\mathrm{d}k$†; some time δt after the application of an electric field $\mathscr{E}$ the electron will have accelerated to a state k_2 having energy E_2. The increase in kinetic energy $\delta E = (E_2 - E_1)$ equals the decrease in electrostatic potential energy:

$$\delta E = e\mathscr{E}v_g\delta t = \frac{e\mathscr{E}}{\hbar}\frac{\mathrm{d}E}{\mathrm{d}k}\delta t$$

so that in the limit

$$\hbar\frac{\mathrm{d}k}{\mathrm{d}t} = e\mathscr{E} \tag{3.9}$$

Equ. 3.9 is the equation of motion for a band electron and does not depend upon any particular relationship between energy and wave-vector (very intense electric fields, far higher than those attainable in metals, can cause Equ. 3.9 to break down; such effects are of importance in semiconductors and insulators).

†For any wave of angular frequency ω, $v_g = \mathrm{d}\omega/\mathrm{d}k$. In our case $E = \hbar\omega$, so that $v_g = (1/\hbar)\,\mathrm{d}E/\mathrm{d}k$.

In the special case of our free-electron band the energy–wavevector relationship is

$$E = \hbar^2 k^2 / 2m$$

so

$$v_g = \hbar k / m$$

and Equ. 3.9 becomes

$$m\dot{v}_g = e\mathscr{E}$$

The dynamics of a free electron in a band are therefore identical with those of a classical particle, as required by the correspondence principle.

Now consider the whole distribution of occupied states: Equ. 3.9 tells us that in a steady electric field every state moves through k-space at the same rate, so that the whole distribution drifts bodily through k-space without change of shape.

So far we have not explained the constancy of current for a given field or allowed for the return to equilibrium after the removal of a field, for it is a matter of everyday experience that the current then stops very quickly indeed. The newly occupied states have higher energy than newly vacated states on the opposite side (Fig. 3.8); if the k-states were exact eigenstates of the system no transitions between them could occur, but a real solid inevitably contains some impurities, and it is (at $T = 0$) these that cause scattering between k-states; this

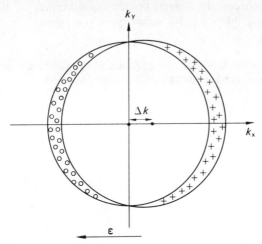

Figure 3.8   The equilibrium and displaced distributions of occupied states in k-space, drawn for a spherical (free-electron) Fermi surface. The displacement has been greatly exaggerated compared with that attainable in any real metal.

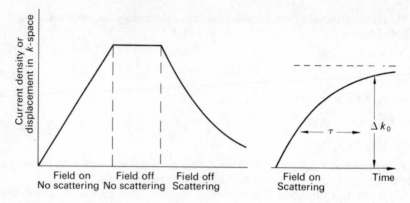

Figure 3.9   The effects of electric field, scattering, and the two combined on an electron distribution in k-space. The current density is proportional to the displacement.

topic is discussed at length in the book in this series by Dugdale, where it is also shown that the thermal vibration of the ions has much the same effect. Although the microscopic details of the scattering processes are complicated this relaxation to equilibrium will have some characteristic time scale τ (Fig. 3.9) which is of the same order as the lifetime of a single electron between scatterings. We can approximate the current decay to a simple exponential of time constant τ, and, much more crudely, imagine that during relaxation the distribution slips bodily back to equilibrium in the same exponential fashion:

$$\Delta k \propto \exp(-t/\tau) \quad \text{or} \quad \dot{\Delta k} = -(1/\tau)\Delta k$$

We can now write down the full equation of motion of the distribution in k-space, with both driving and relaxation terms included:

$$\dot{\Delta k} = \frac{e\mathscr{E}}{\hbar} - \frac{1}{\tau}\Delta k$$

The steady state displacement of the distribution is

$$\Delta k_0 = e\mathscr{E}\tau/\hbar$$

The current density carried by the displaced distribution is

$$J = \sum_{\substack{\text{newly occupied} \\ \text{states}}} ev_g - \sum_{\substack{\text{newly vacated} \\ \text{states}}} ev_g$$

which in the general case of a non-spherical Fermi surface (and a complicated relationship between E and k) will be a difficult sum to evaluate.

However, for a free-electron gas the current can be calculated easily, since for free electrons

$$mv_g = \hbar k$$

and therefore the uniform displacement in k-space is equivalent to a drift velocity for every electron of

$$\Delta v_g = e\mathscr{E}\tau/m$$

The current density is (for N electrons per unit volume)

$$J = Ne\,\Delta v_g = Ne^2\,\mathscr{E}\tau/m$$

and the conductivity

$$\sigma = J/\mathscr{E} = Ne^2\,\tau/m \qquad (3.10)$$

It is most useful to regard this free-electron expression for the conductivity as giving us (by comparison with experiment) some idea of the relaxation time τ and related parameters. First we can see that normal current densities correspond to a very small displacement of the distribution: the electron number density in copper is 10^{23} cm^3; the safe current density for copper wire is about 100 A cm^{-2} and the drift velocity is therefore:

$$\Delta v_g \sim 10^{-2} \text{ cm s}^{-1}$$

which should be compared with a Fermi velocity of 10^8 cm s^{-1}. The corresponding displacement in k-space is about 10^{-2} cm^{-1}, whereas k_{max} is 10^8 cm^{-1}.

The room temperature resistivity of copper is 1.7×10^{-6} Ω cm, so that τ is of order 10^{-13} s, a time which sounds short but which is long enough to allow an electron to travel about 10^{-5} cm between collisions, that is, a few hundred lattice spacings. By cooling metals to liquid helium temperatures so as to eliminate the thermal motion of the ions, and with careful purification, it has been possible to obtain resistivities as low as 10^{-11} Ω cm, which correspond to a mean free path between collisions of about a centimetre – nearly 10^8 lattice spacings. This result is crucial for our description of electrons in solids, since it is essential that the wave packets we have been using should be able to propagate for a large number of wavelengths before being scattered, for otherwise the concept of a wave packet would not be meaningful; furthermore our description of the free-electron gas would not carry conviction if uncertainties in k were an appreciable fraction of k_{max} (the uncertainty principle says that any spatial localization of a particle, as with a finite mean free path, involves loss of information about momentum).

3.6   Failures of the free-electron model

In spite of the successes of the free-electron model, especially for the alkali metals, it has several important shortcomings, some qualitative and some quantitative.

The most striking of the former is its failure to explain how some elemental solids that clearly do have valence electrons, fail to conduct electricity at low temperatures and have apparently no electronic contribution to the specific heat, in spite of reasonable soft X-ray bandwidths for the outer electrons. Si and Ge are the most obvious examples, and are perhaps more surprising because of their metallic appearance.

The transition elements, on the other hand, show no failure of metallic nature, but yield bandwidths too small and densities of states too high to be satisfactorily incorporated in the free-electron fold. In view of their importance in metallurgy, solid state physics and electrical engineering these are failures that cannot be ignored.

Even for some comparatively simple metals like Li, Cu, Ag, Zn and Cd there are some transport properties (thermoelectric powers or Hall coefficients) which show *signs* opposite to those expected on a free-electron model.

It seems very likely that all these effects are due to our neglect of the periodic lattice potential, and we shall now take it into account.

4

Electrons in Metals:
Interaction with the Crystal Lattice

4.1   Consequences of the periodic potential

So far we have entirely ignored the electrostatic interaction between the
conduction electrons and the positively charged ions of the lattice. Our neglect
was not justified and indeed if we compare the   electrostatic interaction
between an electron and a positive ion an atomic spacing apart with the Fermi
energy E_F or E_{max}, we find them to be of similar magnitude. In other words,
typical variations in potential energy for the conduction electrons are the same
size as their kinetic energies, although the free-electron model takes only the
latter into account. It was the relatively recent experimental evidence on the
shape of the Fermi surface of the simple monovalent metals which showed that
the free-electron approximation had much greater validity than originally
supposed. The expected strong electron—ion interaction is weakened by two
effects. First, the ions lie on a regular lattice so that the phenomenon is one of
diffraction by a three-dimensional grating rather than scattering by single ions,
and diffraction occurs for only certain ranges of electron k-vectors†. Second, the
interaction between electron and ion is often far weaker than the straightforward
electrostatic one because of constraints imposed by the exclusion principle which
forces the conduction electrons to avoid the ion core.

Our main object in this chapter is to provide the representations of the
electronic structure that must replace the $E(k)$ relationship, the spherical
constant energy surfaces in k-space and the parabolic density of states curve of
Fig. 3.3 when account is taken of the periodicity of the crystal lattice potential.
The very fact that the system (electrons + lattice) possesses translational
symmetry enables us to make a number of rather general statements about the
form of the eigenstates and the behaviour of E as a function of k but we shall
postpone these until we have introduced a plausible account of the electronic

---

†The smallest k-value for diffraction in the b.c.c. lattice is $2\pi/(a\sqrt{2})$ where a is the lattice
spacing; the value of k_F (= k_{max}) given for Na by the considerations of the previous chapter
is 0·88 times that value.

structure on physical grounds. The essence of this approach is the recognition that at some particular wavelength or k-value an electron travelling through the crystal will be diffracted, and we must therefore expect its energy to be modified from that of a free electron (i.e. one moving in a constant potential) with the same k-value. The criteria for this diffraction will be exactly those for the diffraction of X-rays or beams of electrons directed on to the crystal, that is the familiar Bragg law

$$n\lambda = 2d_{hkl} \sin \theta \tag{4.1}$$

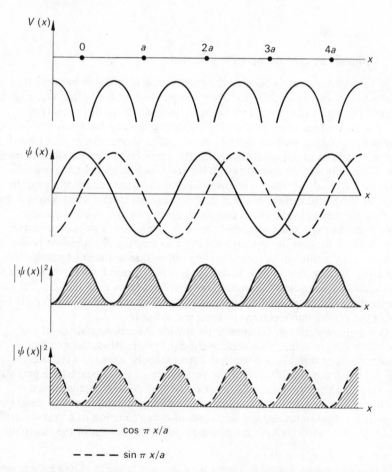

$$\text{——— } \cos \pi x/a$$

$$\text{– – – – } \sin \pi x/a$$

Figure 4.1   Schematic potential for a one-dimensional metal and wavefunctions for the eigenstates of $k_x = \pm\pi/a$ with their charge densities.

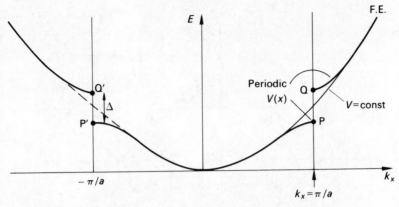

Figure 4.2   Splitting of levels at $k_x = \pm\pi/a$ for the potential of Fig. 4.1.

where λ is the wavelength, d_{hkl} the spacing of the set of crystal planes (hkl), n the order of the diffraction, and θ the angle between the wavevector of the incident radiation and the planes. We shall discuss this condition in the form that is most useful for our present concerns in the next section, after looking at its consequences in general terms.

Let us first consider a one-dimensional situation† where electron states in k-space have k-values only on the k_x axis, that is, our electrons travel along a line of atoms in the x-direction in real space with an interatomic spacing a. The diffraction condition (4.1) reduces to $k_x = \pm n\pi/a$ and we can say that for these k-values the incident wave e^{ikx} is 'reflected' to yield the wave e^{-ikx}. The combinations of these two plane wave travelling in opposite directions leads to standing waves, the two allowed solutions being

$$\psi(x) = \cos \pi x/a \quad \text{and} \quad \psi(x) = \sin \pi x/a$$

for $n = 1$.

The potential and these wavefunctions are sketched in Fig. 4.1. In a uniform potential the cosine and sine wavefunctions would yield the same energy ($\hbar^2 k^2/2m$) but it is clear from the figure that electron density (proportional to $|\psi|^2$) is concentrated at the atom sites, the local potential minima, for the cosine wave, and between the atom sites for the sine wave. This leads in a natural way to an energy difference Δ, and therefore to two states for $k_x = \pi/a$ which will have energies approximately $\hbar^2 (\pi/a)^2/2m + \Delta/2$ and $\hbar^2(\pi/a)^2/2m - \Delta/2$ (Fig. 4.2). (The interaction of the electron states with the crystal lattice is rather more complicated than the electrostatic coupling between a simple sine or cosine

†The general three-dimensional situations is treated in § 4.4.

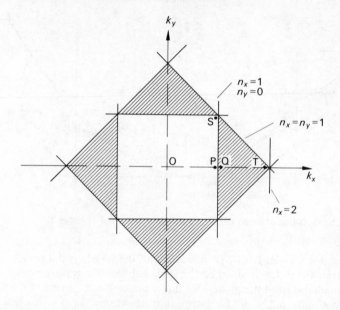

Figure 4.3   Lines in k-space satisfying the diffraction condition.

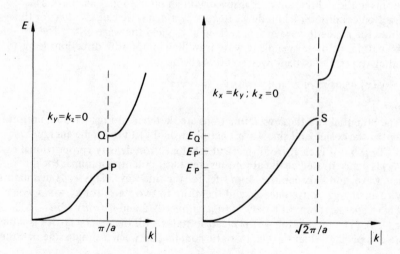

Figure 4.4   Energy as a function of $|k|$ in the k_x and k_{xy} directions in k-space for a simple-cubic lattice.

wave with the potential shown, and while there are two distinct states with an energy gap between them the labelling of the upper and lower states is not always simple. Very close to a particular nucleus we expect the wavefunction to take the form of an atomic wavefunction and therefore to have either a cusp or node.)

If the influence of the periodic part of the potential on the allowed states is small we expect the $E-k_x$ relationship to remain fairly close to that for free electrons except when $k \sim k_{\text{diff}}$. Furthermore, since at the diffraction condition we have standing waves which cannot propagate through the crystal, the group velocity at $k_x = \pm\pi/a$ will vanish and since (see p. 56) the group velocity is $(1/\hbar) \, dE/dk$ we can draw $E(k)$ as in Fig. 4.2 with the slope becoming zero where the energy gap appears. Notice that for the periodic potential there exist *no* allowed states with energies between P and Q.

The extension of these arguments to a three-dimensional simple-cubic lattice is straightforward. The diffraction condition is satisfied whenever the k_x or k_y or k_z component of k takes the value $(n\pi/a)$ and in Fig. 4.3 we show the locus of such points in the $k_x k_y$ plane, and in Fig. 4.4 the behaviour of $E(k)$ for two different directions in k-space. For some values of E we clearly have values of $| k |$ that vary with the direction of k and the constant energy surfaces will be deformed from spheres as shown in Fig. 4.5. For energy values greater than E_P the range Δ of forbidden states will modify the situation. Thus at energy E_H there will be no allowed states along the line PH and the density of allowed

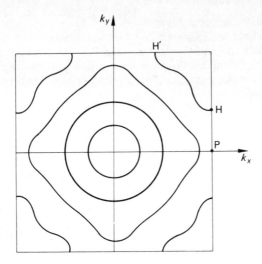

Figure 4.5   Constant energy surfaces in the $k_x k_y$ plane in k-space for a simple-cubic lattice.

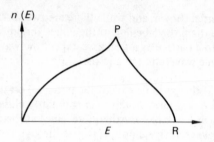

Figure 4.6   The density of states for the constant energy surfaces of Fig. 4.5.

states at E_H will be contributed to only by states lying in curved segments of the constant energy surface like HH′. The density of allowed states per unit energy will therefore be less than the free-electron value for the same energy, and will *decrease* with increasing energy as the states for which k has too small a component along k_x or k_y or k_z to satisfy the diffraction condition are exhausted; the density of allowed states $n(E)$ as a function of energy will take the form shown in Fig. 4.6. (The cusp on $n(E)$ at the energy E_P where diffraction first takes place is a consequence of the fact that (see p. 50) the expression for the density of states is

$$n(E) = \frac{1}{8\pi^3} \int \frac{dS}{\mathrm{grad}_k E}$$

and dE/dk_x tends to zero as k_x approaches P).

Thus we now have in Figures 4.5 and 4.6 the representations of the electronic structure that must replace those of Fig. 3.3. Notice that the longest k-vector for which diffraction does not occur is that of the state just below that with $k_x = k_y = k_z = \pi/a$, and this will be R, the highest energy state in the zone (see Figures 4.6 and 4.7).

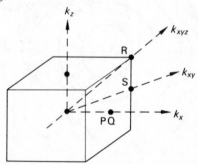

Figure 4.7   The volume (or zone) in k-space bounded by planes satisfying the diffraction condition.

4.2   Diffraction by a periodic potential

We give here a rather more formal and general account of the conditions for diffraction of the electrons by the periodic lattice, introducing the reciprocal lattice often used by crystallographers. Most of the discussions in later sections of this chapter can be followed however omitting this section.

We start with the plane wave eigenstates of Chapter 3 and allow them to interact weakly with a regular array of positive ions; the problem thus parallels that of X-ray diffraction considered by Brown and Forsyth in book 3 of this series (*The Crystal Structure of Solids*). For simplicity we consider a crystal with one atom per unit cell on the points of a lattice defined by the three primitive lattice translation vectors a, b, c; a given site is then defined by the lattice translation vector

$$T = n_1 a + n_2 b + n_3 c \tag{4.2}$$

Each ion will scatter a spherical wave from the incident beam and because energy is conserved the frequency and wavenumber of the scattered wave will be the same as that of the incident wave; the phase will depend on the incident phase at the scattering site. Consider the scattered waves from two ions A and B (Fig. 4.8); they will add constructively on a wavefront of a diffracted wave if the

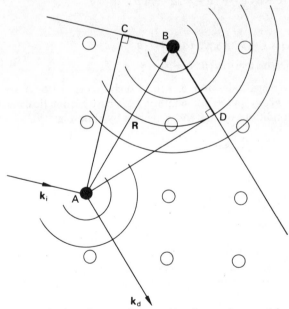

Figure 4.8   Constructive interference of two scattered waves in a crystal.

path difference CBD is an integer number of wavelengths (notice that this arguement is identical to the elementary theory of an optical diffraction grating). By vector geometry $CB = k_i \cdot R / |k_i|$ and $BD = -k_d \cdot R / |k_d|$; the wavelength is $2\pi / |k_i| \; (\equiv 2\pi / |k_d|)$. Therefore the condition for the two scattered waves to add constructively is:

$$(k_i - k_d) \cdot R = 2m\pi \quad (m \text{ integer})$$

It is useful to define a scattering (wave) vector $K = (k_i - k_d)$. For all the ions in the crystal to add constructively we require that R takes all allowed values of T, and $(K \cdot T)/2\pi$ shall always be an integer. Denote the values of K that satisfy this condition by G, and we can write the equation as

$$\exp(iG \cdot T) = 1 \tag{4.3}$$

G clearly has the dimensions of wave-vector, and can be resolved into three components. A convenient way to solve Equ. 4.3 is to split G into components α, β and γ that are perpendicular to b and c, c and a, and a and b respectively; that is

$$G = \nu_1 \alpha + \nu_2 \beta + \nu_3 \gamma \tag{4.4}$$

with

$$\alpha \cdot b = \alpha \cdot c = \beta \cdot c = \beta \cdot a = \gamma \cdot a = \gamma \cdot b = 0.$$

Then

$$G \cdot T = (\nu_1 \alpha + \nu_2 \beta + \nu_3 \gamma) \cdot (n_1 a + n_2 b + n_3 c)$$

$$= \nu_1 n_1 \alpha \cdot a + \nu_2 n_2 \beta \cdot b + \nu_3 n_3 \gamma \cdot c$$

If we choose the magnitudes of α, β, γ to be such that $\alpha \cdot a = \beta \cdot b = \gamma \cdot c = 2\pi$ and also require ν_1, ν_2 and ν_3 to be integers, we have automatically satisfied Equ. 4.3. Our conditions on α, β and γ are met by choosing:

$$\alpha = 2\pi \frac{b \wedge c}{a \cdot b \wedge c}, \quad \beta = 2\pi \frac{c \wedge a}{a \cdot b \wedge c}, \quad \gamma = 2\pi \frac{a \wedge b}{a \cdot b \wedge c} \tag{4.5}$$

The allowed values G for which the scattering vector gives a diffracted wave define a lattice (compare Equ. 4.2 with Equ. 4.4), but one in k-space rather than in real space; it is known as the *reciprocal lattice* (crystallographers sometimes define the reciprocal lattice without the factor of 2π in Equ. 4.5). The G's are therefore the translation vectors of the reciprocal lattice, and they have a simple physical interpretation. Consider a wave $\exp(iG \cdot r)$, which has wave-vector equal to some particular reciprocal lattice vector. It has the same phase at all lattice points (from Equ. 4.3), so that it must have wavefronts, which are planes of equal phase, that pass through all lattice points. The wavefronts of this wave are therefore just a set of crystal planes, and it is not hard to show that the Miller indices h, k and l have the same ratio as v_1, v_2 and v_3, i.e. $h:k:l$ is as $v_1:v_2:v_3$.

Another important property that follows from Equ. 4.3 is that the reciprocal lattice vectors represent all the spatial frequencies of the lattice. If any physical property of the crystal which has the full translational symmetry of the lattice, for example the charge density or the electrostatic potential of the ions, is Fourier analysed into sine waves, as $\sin q \cdot r$ or $\exp(iq \cdot r)$, the only non-zero components are those with q equal to a reciprocal lattice vector.

The simplest way of expressing the diffraction condition is to specify the locus in k-space of those k-vectors for which it will be satisfied. This is

$$K = k_i - k_d = G \tag{4.6}$$

and

$$|k_i| = |k_d| \tag{4.7}$$

Equ. 4.6 yields

$$(k_i - k_d)^2 = |k_i|^2 - 2k_i k_d + |k_d|^2 = |G|^2$$

or

$$2|k_i|^2 - 2k_i \cdot (k_i - G) = |G|^2$$

giving

$$2k_i \cdot G = |G|^2 \tag{4.8}$$

as the Bragg condition for 'reflection' from a set of lattice planes†

Thus a wave-vector for diffraction k_i has a component parallel to some reciprocal lattice vector G equal to half the length of G. The loci in k-space of the tips of such k_i are the planes normal to vectors G that bisect G.

4.3   Brillouin zones and the nearly-free-electron model

The conduction electron states with lowest energy lie at the centre of k-space, and it is not until some minimum wavevector k_{min} (equal to one half the smallest reciprocal lattice vector) has been reached that Bragg reflection can occur. We have seen that within the free-electron model the energy contours are spherical, and that the radius of the Fermi surface is easy to calculate. The significant question now is whether the Fermi surface has radius k_F considerably

---

† As expressed in Equ. 4.8 the order of reflection n in the traditional statement of Bragg's Law $n\lambda = 2d_{hkl} \sin \theta$ is incorporated in v_1, v_2 and v_3 so that unlike true Miller indices these indices may possess a common integer factor.

less than $|\,k_{\min}\,|$, or if it is larger than $|\,k_{\min}\,|$; in the former case none of the occupied states satisfy the Bragg condition, no diffraction can occur, and to a good approximation the conduction electrons are unaware of the positive ions. On the other hand, with a larger Fermi surface we will have to look in some detail at the way in which diffraction modifies the behaviour of the conduction electrons.

k-space is dissected by the planes on which the Bragg condition (Equ. 4.8) is satisfied into segments called Brillouin zones; the first Brillouin zone is that part of k-space surrounding the origin and bounded by the first set of planes; the nth Brillouin zone is reached after crossing $(n-1)$ planes and is bounded by the nth plane (Fig. 4.9). Consider the prescription for obtaining the first Brillouin zone: we have a lattice (in k-space, but the argument that follows holds just as well for a real space lattice) and we divide it up by drawing the perpendicular bisectors between each lattice point, the bisectors intersect and form cells around each lattice point. There is one cell to each lattice point and the cells are space-filling, therefore these cells, (known as Wigner—Seitz cells in the real lattice and Brillouin zones in wave-vector space) are primitive unit cells

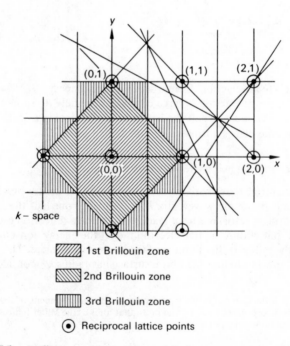

Figure 4.9   Brillouin zones in two dimensions for a square lattice. The x and y axes are not themselves Brillouin zone boundaries.

of the corresponding lattices, the crystal lattice for the former and the reciprocal lattice for the latter, and must have the same volume as the more usual choice of primitive unit cell, which is a parallelepiped bounded by the three primitive translation vectors, a, b and c or α, β and γ. It is a standard result of vector geometry that this volume is $(a \cdot b \wedge c)$ or $(\alpha \cdot \beta \wedge \gamma)$ as the case may be.

Secondly, each Brillouin zone has the same volume as the first, as can be seen (Fig. 4.9) by assembling the component parts of a zone within the first zone. Notice that the translations that are required for the re-mapping process are reciprocal lattice vectors.

We are now in a position to compare the capacity of the Brillouin zones with the number of states to be occupied by the conduction electrons. The Brillouin zone volume Ω_{BZ} from Equ. 4.5 is given by

$$\Omega_{BZ} = (\alpha \cdot \beta \wedge \gamma) = \frac{(2\pi)^3}{(a \cdot b \wedge c)^3} \cdot (b \wedge c) \cdot (c \wedge a) \wedge (a \wedge b)$$

$$= \frac{2\pi^3}{a \cdot b \wedge c} = \frac{2\pi^3}{V} \tag{4.9}$$

where V is the real space unit cell volume. (For the simple cube this is obvious from Fig. 4.7.) We have seen that in k-space each k-state has associated with it a volume $(2\pi)^3/L^3$ where, in order to obtain the correct boundary conditions, we considered a macroscopic cube of side L. The number of states in the zone is therefore L^3/V, and we arrive at the important and simple result that

The number of k-states in a Brillouin zone equals the number of real space unit cells in the macroscopic sample.

The choice of a cube as the macroscopic volume of our specimen was one of convenience and the result holds good for any shape, provided that the sample is large enough for surface effects to be unimportant.

Many metallic elements have a structure in which there is one atom per unit cell, for these the occupied volume of k-space will equal $\frac{1}{2}\Omega_{BZ}$ for the monovalent metals (remembering that each k-state accommodates two electrons), Ω_{BZ} for the divalent, and so on. This occupied volume is incompressible and in the spirit of the free-electron model it will endeavour to retain a spherical shape.

No elemental metal has simple-cubic structure, but the simplicity of this structure makes it a useful case to consider. The primitive translation vectors are $a = ai$, $b = aj$, $c = ak$ where a is the lattice spacing and (i, j, k) a set of Cartesian unit vectors. The reciprocal lattice vectors are $\alpha = (2\pi/a)i^*$, $\beta = (2\pi/b)j^*$, $\gamma = (2\pi/c)k^*$ where (i^*, j^*, k^*) are unit vectors in k-space and parallel to (i, j, k). The first Brillouin zone (Fig. 4.7) is therefore a cube, and for Bragg reflection to occur $k_{min} = \pi/a$; the free-electron Fermi sphere has volume $\frac{1}{2}(2\pi/a)^3$, and therefore radius $k_F = \sqrt{(3/\pi)} \, \pi/a = 0.98 \, k_{min}$. A monovalent simple-cubic metal would have a free-electron Fermi surface that just fits within the first Brillouin zone.

In real monovalent metals the fit is not so tight; for example the alkali metals have a body-centred cubic (b.c.c.) structure. Although it is generally easier to think about the b.c.c. structure in terms of a cubic unit cell of side a which contains *two* lattice points, a truly primitive unit cell is outlined by primitive translation vectors

$$a = (a/2)(i + j + k)$$
$$b = (a/2)(i + j - k)$$
$$c = (a/2)(-i + j + k)$$

and the corresponding primitive reciprocal lattice vectors are

$$\alpha = (2\pi/a)(i^* + k^*)$$
$$\beta = (2\pi/a)(j^* - k^*)$$
$$\gamma = (2\pi/a)(i^* + j^*)$$

Incidentally, these primitive translation vectors describe a face-centred cubic (f.c.c.) lattice based on a cube of side $4\pi/a$, so that the f.c.c. and b.c.c. lattices are reciprocals of one another. The shortest reciprocal lattice vectors have length $|\alpha| = |\beta| = |\gamma| = \sqrt{2}(2\pi/a) = 2k_{min}$. The real space primitive unit cells volume V is half that of the cubic unit cell, therefore $\Omega_{BZ} = (2\pi)^3/\frac{1}{2}a^3$, consequently the monovalent Fermi surface has volume $(2\pi/a)^3$ and a free-electron sphere has radius $k_F = (2\pi/a)(3/4\pi)^{1/3}$ that is, $k_F = 0.88 \, k_{min}$. Monovalent f.c.c. metals have, when a similar comparison is made, $k_F = 0.91 \, k_{min}$.

Using the traditional crystallographer's cubic unit cell and related notation for the body-centred cube we would say that the minimum k-value for which diffraction takes place is that corresponding to diffraction by the (110) planes, and the Brillouin zone (Fig. A.4.1, p. 104) is bounded by planes where the diffraction condition is satisfied for the *family* of planes {110}†. The f.c.c. situation is rather more complicated when we use the cubic cell and corresponding notation; although the shortest k-vector for which diffraction occurs is that corresponding to (111) plane diffraction, the shortest k-vector for diffraction of electrons propagating along the cube axes is $(\pi/2a)$, diffraction being caused by planes of the family {200}. Thus the Brillouin zone (see Fig. A.4.1, p. 104) has boundary planes of two types, square faces where {200} type diffraction occurs and hexagonal faces where {111} diffraction occurs. We may still express the number of allowed states in the Brillouin zone in terms of the number of *atoms* in the crystal for the b.c.c. and f.c.c. crystals, since the {110}

---

†With the conventional cubic unit cell the {100} planes are more widely spaced than the {110} planes, or equivalently the corresponding reciprocal lattice vector is shorter. The crystallographer describes the absence of diffraction by these planes as a vanishing of the 'structure factor.' From our point of view it is simply that they do not correspond to a true lattice periodicity.

Brillouin zone has twice the volume of the basic {100} cubic Brillouin zone and therefore holds twice as many states as there are cubic unit cells (each having two atoms) in the crystal. For the f.c.c. crystal with four atoms per unit cell the same argument holds since the {111} {200} Brillouin zone has four times the volume of the {100} Brillouin zone.

So far we have been able to infer that states well away from Brillouin zone boundaries are unaffected by the lattice, and that those actually on the boundaries become standing waves, but what of states close to the boundaries? They correspond to a condition in which a diffracted wave nearly builds up, but the contribution from ions spaced far apart causes cancellation. We can expect the propagation of such a wave packet to be seriously affected, and a modulation of the charge density to be introduced. The energy band will therefore deviate smoothly from the free-electron curve in the manner indicated in Fig. 4.4 and correspondingly we can anticipate that the energy contours will bend smoothly in to meet the Brillouin zone boundaries at right angles.

No matter how weak the interaction is between electrons and ions, the periodicity of the lattice ensures that Bragg reflection of states at the Brillouin zone boundaries takes place, but the size of the deviations elsewhere from free-electron behaviour will depend upon the strength of the interaction. The nearly-free-electron model (NFEM) of a metal was developed for the case in which the interaction is weak in the following sense: deviations from the free-electron band are everywhere small compared with the kinetic energy, or equivalently: the Fermi surface remains roughly spherical.

The consequences of variations in the strength of the electron lattice interaction can be very striking. Thus if we go back to Fig. 4.7 and draw E as a

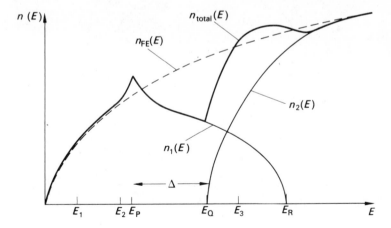

Figure 4.10 Density of states for overlapping bands.

function of $|k|$ for the directions labelled k_x and k_{xyz} (instead of k_x and k_{xy} as in Fig. 4.4) we will expect (Fig. 4.10), if the energy gap Δ across the zone face at P is small, that E_R will lie as shown above E_Q; after all, $E_R \cong (\hbar^2/2m)$ $(3\pi^2/a^2) - (\Delta/2)$ and $E_Q \cong (\hbar^2/2m)(\pi/a)^2 + (\Delta/2)$. The resultant density of states curve is the sum of those inside and outside the Brillouin zone, and (Fig. 4.10) will not be so very different from the free-electron density $n_{fe}(E)$. The resultant constant energy surfaces for $E_Q < E < E_R$ will have two sets of sheets, those (see Fig. 4.11) inside the Brillouin zone and those outside.

On the other hand a large energy gap (compared with $(\hbar^2/2m)(\pi/a)^2$) will place all states outside the zone at higher energy than all those inside it (see Fig. 4.12), and we shall examine the consequences in Section 5.1.

From the discussions of this section it should be clear that the free-electron model works well for some monovalent metals because for small Brillouin zone energy gaps the Fermi surface can remain spherical, and quite large gaps are required if that surface is to touch the zone faces in b.c.c. or f.c.c. structures.

For any crystal structure the first Brillouin zone is a polyhedron and an inscribed sphere must necessarily correspond to less than two electrons per atom, so that Bragg reflection must occur for some of the conduction electrons in a *polyvalent* metal. A nearly-free-electron model and an approximately spherical Fermi surface will then be appropriate only for Brillouin zone energy gaps quite small compared with the Fermi energy.

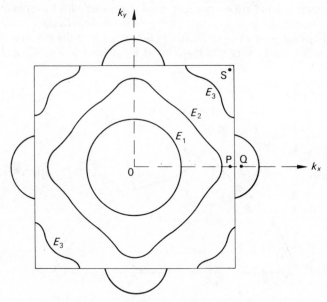

Figure 4.11   Constant energy surfaces for the situation of Fig. 4.10 (small gaps).

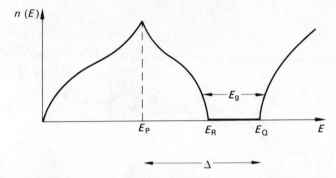

Figure 4.12   The density of states for large energy gaps.

4.4   Electron eigenstates in a crystal

We have seen that free electrons, that is electrons in a constant potential, have eigenstates of the form $e^{ik \cdot r}$. The charge density $| \psi |^2$ is therefore spatially uniform, which satisfactorily reflects the translational *invariance* of such a potential: all positions within it are equivalent.

If we now think about a crystal, translational invariance has been lost, but it is replaced by translational *symmetry*. Every physical observable must reflect that symmetry so as to ensure that all unit cells are identical. So, for the charge density we must now have

$$| \psi(r + T) |^2 = | \psi(r) |^2 \tag{4.10}$$

where T is a lattice translation vector.

It is easy to see that functions of the form:

$$\psi_k(r) = u_k(r) \exp(ik \cdot r) \tag{4.11}$$

satisfy Equ. 4.10 provided that $u_k(r)$ has the lattice periodicity:

$$u_k(r + T) = u_k(r)$$

These are known as Bloch functions†, and can be thought of as plane waves ($e^{ik \cdot r}$) modulated by a periodic function $u_k(r)$. The charge density and so on repeat from unit cell to unit cell, but the quantum mechanical phase (which is not a physical observable) advances by $e^{ik \cdot T}$ between two unit cells separated by T.

If the atoms are far apart compared with the typical extent of electronic wavefunctions (which is the tight-binding situation, and is discussed in more

†Formal proofs that the allowed eigenstates in a periodic potential are Bloch functions will be found in the text books recommended on p. 134.

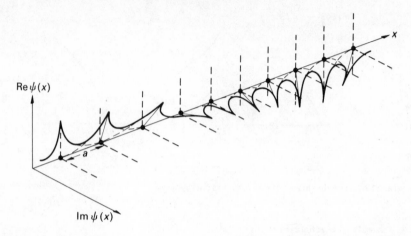

Figure 4.13   A one-dimensional Bloch function of small overlap (tight-binding) character for $\phi_{at} = 1s$ and $k_x = \pi/8a$.

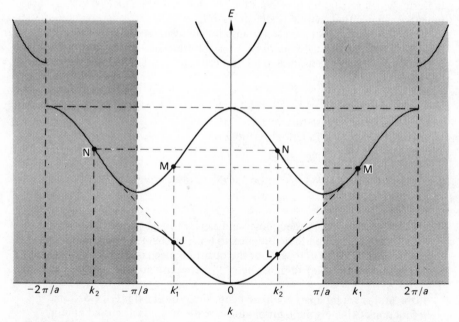

Figure 4.14   Labelling of k states; in the reduced zone representation the states M and N are plotted within the first Brillouin zone at k_1' and k_2'. They are different states from J and L.

detail in § 4.5 below), then we can expect $u_k(r)$ to resemble closely atomic wavefunctions. The phase factor $e^{ik.r}$ ensures that the crystal wavefunction follows the Bloch prescription, but has very little effect upon the charge distribution (i.e. on $|\psi(r)|^2$) if the distance between atoms is large (Fig. 4.13). It is equally instructive to see the way in which Bloch states emerge from the free-electron model of a crystal. In that case, we start with plane wave eigenstates $e^{ik.r}$, and allow them to interact with the regular three-dimensional array of positive ions. We have seen that if we start with a plane wave of wave-vector k, the wavelets scattered by each ion contribute to plane waves of wave-vector $k + G$, where G is any of the reciprocal lattice vectors. It is only in the special case that one of the secondary waves has the same energy as the original wave (i.e. $|k| = |k + G|$), that diffraction occurs, and an incident wave k can be diffracted and emerge from the crystal with wave-vector $k + G$ (or vice versa). Otherwise, no diffracted beam can get out of the crystal, but still the secondary waves within the crystal are important, the more so the nearer the diffraction condition is to being satisfied. The contrast is best brought out in the language of perturbation theory. If none of the original plane wave states of wave-vector $k + G$ has energy close to that of the original state k, compared with the size U of the interaction, then the next order approximation to the wavefunction is

$$\psi_k(r) = \exp(ik.r) + \sum_G \frac{U}{E_{k+G} - E_k} \exp[i(k + G).r] \qquad (4.12)$$

where E_k and E_{k+G} are the energies of the original plane wave states in the absence of the periodic potential. On the other hand if, for some particular $G, E_k \sim E_{k+G}$ the coefficient of $e^{i(k+G).r}$ becomes large, and degenerate perturbation theory must be used instead.

In any event the interaction of the crystal potential, with periodicity described by reciprocal lattice vectors G, with a free-electron plane wave state is always to give a new eigenstate of the form:

$$\psi_k(r) = \exp(ik.r) + \sum_G A_G \exp[i(k + G).r] \qquad (4.13)$$

where the coefficients A_G are small provided that the diffraction condition $|k| = |k + G|$ is far from being satisfied.

We can see straight away that such states do satisfy the Bloch criterion for we can write

$$\psi_k(r) = \exp(ik.r) \left[1 + \sum_G A_G \exp(iG.r) \right]$$

and the second term certainly has the periodicity of the lattice:

$$1 + \sum_G A_G \exp[iG.(r + T)] = 1 + \sum_G A_G \exp(iG.r)$$

since by the definition of G we have $\exp(iG \cdot T) = 1$ (Equ. 4.3).

There is one case where this periodic part of the wavefunction has a simple form. Consider some particular G: a plane wave of wave-vector $G/2$ gets diffracted into a plane wave of wavevector $-G/2$, and vice versa. Consequently the eigenstates must consist of a coherent mixture of these two plane waves, so that (provided all the other A_G are small):

$$\psi_{G/2}(r) = \exp(iG \cdot r/2) + C \exp(-iG \cdot r/2)$$

What is C? The original plane waves are on an equal footing, since each is diffracted into the other, so therefore $|C| = 1$. What about the phase of C? Here we can use another symmetry argument, for not only must the charge density show the periodicity of the crystal, but there is also (for the simple crystals that we are considering) inversion symmetry about any lattice point: the charge density must be the same at r and $-r$. This condition restricts the choice of C to ± 1, and the two states are:

$$\psi_+ = \exp(iG \cdot r/2) + \exp(-iG \cdot r/2) = 2 \cos(G \cdot r/2)$$

giving $|\psi_+|^2 = \frac{1}{2}[1 + \cos(G \cdot r)]$ and

$$\psi_- = \exp(iG \cdot r/2) - \exp(-iG \cdot r/2) = 2 \sin(G \cdot r/2)$$

giving $|\psi_-|^2 = \frac{1}{2}[1 - \cos(G \cdot r)]$.

These states and their associated charge densities were illustrated in Fig. 4.1 for the case of a simple-cubic lattice and G, the reciprocal lattice vector, $(2\pi/a, 0, 0)$. As we saw in §4.1 because the two states distribute the electronic charge differently in the electrostatic potential, necessarily they will have different total energies.

If we consider the time dependence of the wavefunction

$$\psi_\pm = \left\{ \begin{matrix} \cos(G \cdot r/2) \\ \sin(G \cdot r/2) \end{matrix} \right\} \exp\left(-i\frac{E_\pm t}{\hbar}\right)$$

it is apparent that these two states are standing waves, in contrast to the travelling wave form of a general Bloch state. Thus they have zero group velocity, which requires that at this point in the band dE/dk must be zero too. Consequently, the shape of the electron energy band must now be as in Fig. 4.2, where interaction with the periodic potential of the lattice has introduced band gaps at the particular values of $k = G/2$.

We have seen above how the same ideas carry over into thinking about the shape of energy bands and energy contours in three-dimensional k-space. Wherever the diffraction conditions

$$k_1 = k_2 + G; \qquad |k_1| = |k_2|$$

are satisfied, a band gap appears; at these points in k-space the energy jumps discontinuously.

On a zone boundary, the k vector must be equal to $k_\parallel + k_\perp$, with $k_\parallel = G/2$ where $k_\perp$ measures the component of wave-vector perpendicular to G. The lattice potential mixes plane wave of this wave-vector with those of wave-vector $(-G/2) + k_\perp$, so that the new eigenstates, with their time dependence included, are

$$\psi_\pm = \{\exp[i(G/2 + k_\perp).r] \pm \exp[i(-G/2 + k_\perp).r]\}\exp(-iE_\pm t/\hbar)$$

$$= \exp[i(k_\perp.r - E_\pm t/\hbar)]\left\{\begin{array}{c} \cos \\ \sin \end{array}\right\}(G.r/2)$$

These states are travelling waves in the $k_\perp$ direction, but standing waves in the G direction. In three dimensions the group velocity is $(1/\hbar)\nabla_k E$ and its component in the G direction is $\partial E/\partial k_\parallel$, which must be zero for these states. Therefore, in any cross section of k-space, as a function of $k_\parallel$ the energy must have an extremum at the zone boundary (Fig. 4.4). Equivalently, the contours of constant energy must approach the zone boundaries normally (Fig. 4.5).

4.4.1   Extended, reduced and repeated zone schemes

In the free-electron model of electrons in a crystal the wave-vector label is unambiguous, for it is simply the wave-vector of the plane wave eigenstate. The labelling of Bloch states is not so clear, for example we have seen that the exact eigenstates can be written as a (potentially infinite) sum over plane waves

$$\psi_k(r) = \exp(ik.r) + \sum_G A_G \exp[i(k + G).r] \tag{4.13}$$

With which wave-vector, k or one of the $(k + G)$'s, should the state be labelled? The answer is that it really does not matter, as long as a consistent scheme is used, and three such schemes are in common use.

(i)   Extended zone scheme

Label the state with the wave-vector of that plane wave that is dominant in the sum; physically this means staying as close as possible to the original free-electron scheme (Fig. 4.2). It should be noted, however, that a slight ambiguity in labelling exists since P and P' are the same state (which is therefore shown twice); they both contain equal amounts of $e^{iG.r/2}$ and $e^{-iG.r/2}$. In the same way Q and Q' are the same state.

In this representation k-space is infinite and if the energy gap is small (see Fig. 4.11) a surface of constant energy (E_3) with E rather greater than the top of that gap will approximate to a sphere except where it intersects the planes of the Brillouin zone. Comparisons with a really free-electron model are then easily made. In Section 4.3 we defined the states reached by crossing $(n - 1)$ planes satisfying the diffraction condition as lying in the nth Brillouin zone and in the extended zone scheme the nth Brillouin zone is a composite of segments of k-space between the $(n - 1)$th plane and the nth plane.

When we go outside the first Brillouin zone there are large numbers of choices of volumes of k-space bounded by planes of energy discontinuity (Fig. 4.9). These are not necessarily true Brillouin zones. It is useful, however, to use some such 'zones' (*Jones zones*) in discussions of the stabilities of crystal structures. The Jones zones chosen are those made up by planes for which strong diffraction occurs and which lie close to the Fermi surface in its extended representation. This has sometimes led to the choice of Jones zones whose volume is not an integral multiple of that of the first Brillouin zone. (See Appendix A.4.1, p. 104 and Fig. A.4.2.)

(ii) Reduced zone scheme

When we write a Bloch function in the form of Equ. 4.13 it is natural to think of it as derived from the plane wave state of wave-vector k. However, we can also write this eigenstate (of energy E_k) in the form

$$\psi_k(r) = u_k(r)\exp(ik . r)$$

(where $u_k(r) \equiv u_k(r + T)$), so that we can describe it by a wave-vector k' such that $k' = k + G$ and

$$\psi_{k'}(r) = v_k(r)\exp(ik' . r)$$

which is also a Bloch function since

$$v_k(r) = u_k(r)\exp(iG . r)$$

and has the property $u_k(r) \equiv u_k(r + T)$ since $e^{iG . T} = 1$.

Thus we could choose to label an eigenstate k_1 of energy E_{k_1} (Fig. 4.14) with wave-vector k'_1 ($=k_1 - (2\pi/a)$) ar•d the different eigenstate k_2 with wave-vector k'_2 ($=k_2 + (2\pi/a)$). For nearly-free-electron states this might seem perverse (since the energy of these two states must be very close to the free-electron energy of plane wave states $e^{ik_1 r}$ or $e^{ik_2 r}$ which is $\hbar^2 k^2/2m$) but it has the advantage of enabling us to plot the energies of all allowed states along the k_x-axis as multivalued functions of k_x in the range $-\pi/a < k_x < + \pi/a$, or for three dimensions as multivalued functions of a k that always lies inside the first Brillouin zone. (For the nearly-free-electron situation it is helpful to use an index n to indicate the Brillouin zone within which a plane wave state of corresponding energy would be found.) This is the reduced zone representation.

In the tight-binding situation (§4.5) the eigenstates no longer bear any resemblance at all to a single plane wave, or equivalently several of the coefficients A_G in the sum of Equ. 4.13 are large. The reduced zone scheme is

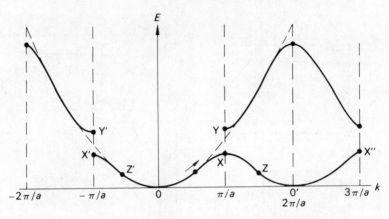

Figure 4.15   The construction of the repeated zone scheme.

then the natural one to choose, because it avoids focusing on any one plane
wave component of the wavefunction.

(iii) The repeated zone scheme

When we are considering the response of an electron in a state k to an applied
electric field $\mathscr{E}$ (see §3.5) it is clear that as it gains energy it moves through states
given by the $E - k$ curves we have shown. In the special situation of a single
electron in a perfect crystal (i.e. no scattering) the k-vector will move along the
$E - k$ curve at a constant rate $\hbar\dot{k} = e\mathscr{E}$ (Equ. 3.9). The diffraction condition will
be satisfied when k lies on the Brillouin zone boundary at the point X in
Fig. 4.15, and because of Bragg 'reflection' we may think of the electron as
'reappearing' at point X'. We have already emphasized that this is the same state
as X and therefore no discontinuity in behaviour occurs. As k then continues to
increase the state occupied will move down the curve X'Z'. In view of the
equivalence we have already pointed out between k and $k + G$ (or in this case
k_x and $k_x + (2\pi/a)$) it is more natural to think of the occupied state as traversing
the path O X Z O' in k-space. A higher band of levels can clearly be treated in
the same way and in one dimension the repeated zone scheme is shown in
Fig. 4.15. It should be noted that in the repeated zone scheme every eigenstate is
shown many times, but it can be *occupied* by only one electron of each spin
direction.

This approach is particularly useful when considering the sequence of allowed
states for an electron constrained by a magnetic field† to move on the Fermi

†In a magnetic field electron states move in k-space along energy contours that are in a
plane normal to the field.

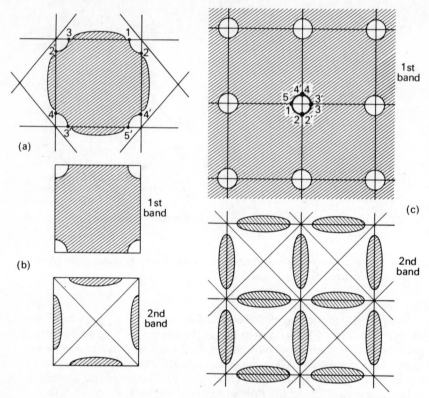

Figure 4.16   Fermi surfaces in (a) extended (b) reduced (c) reduced and repeated zones.

surface when the latter intersects Brillouin zone boundaries. Thus in Fig. 4.16a we show the Fermi surface in the extended zone scheme for such a situation with an almost spherical Fermi surface (i.e. small energy gaps). In Fig. 4.16b the two sheets of the Fermi surface are shown in the reduced zone scheme and in Fig. 4.16c these are shown for the repeated zone scheme. Consider an electron constrained by an applied magnetic field to move on the Fermi surface along the path $1 \rightarrow 2$ in Fig. 4.16a; at 2 it will undergo Bragg reflection to $2'$ and then move along $2' \rightarrow 3$, undergoing reflection to $3'$, moving along $3' \rightarrow 4$, being reflected to $4'$, moving along $4' \rightarrow 5$ and finally being Bragg reflected back to its original state 1. In the repeated zone scheme this motion is represented more smoothly as the path shown in Fig. 4.16c. These 'orbits' on sheets of Fermi surface can be explored by de Haas–van Alphen effect measurements and other techniques to derive information on the connectivity and shape of Fermi surfaces. Notice also that what was a clockwise sense of rotation in Fig. 4.16a becomes anticlockwise in Fig. 4.16c; consideration of the velocity vector, which is normal to the energy contour, shows that the real space motion is indeed anticlockwise, whereas that for a free electron would have been clockwise. This change of sense of rotation is characteristic of sheets of Fermi surface that in the repeated zone scheme enclose empty states ('hole' Fermi surfaces), and it accounts for the anomalous sign of the Hall coefficient in many metals.

4.5   The tight-binding model for d-states

In the previous sections of this chapter we have been considering the modification of free-electron gas eigenstates by the introduction of a small periodic term in the potential. There has been, implicitly, a sharp distinction between conduction electrons in these states and ion core electrons whose states are essentially unchanged from those in corresponding free atoms. This distinction is easily made for Na where the 3s-electron of the atom becomes the 'free' conduction electron in the solid, while the ten electrons of the $(1s)^2 (2s)^2 (2p^6)$ closed shells are little affected by the condensation. (It would not be true to say that the latter are totally unaffected since the redistribution of 3s-electron charge cloud will very slightly modify the Coulomb interactions of the core electrons with the nucleus and with one another.)

There are elements whose chemical behaviour displays a different situation — the transition elements. In these the atomic configuration includes a part-filled d-shell, 3d in the first row transition elements (Sc to Ni), 4d in the second row (Y to Pd) and 5d in the third row (La to Pt). In chemical combination these elements ionize with a variable loss of electrons from the d-shell, sometimes none, sometimes one or two; transition elements near the end of a row sometimes gain a d-electron or two and fill the d-shell. In the transition metals these d-electrons are much too easily removed to allow them to be regarded as well localized components of an ion core, but on the other hand they are not as free as s-electrons, for a transition element never loses more than a couple of d-electrons when it ionizes. A free-electron model would therefore be inappropriate for describing these d-electrons, and instead we approach the problem by looking at a regular array of atomic orbitals in which electrons are fairly well localized; this point of view is known as the tight-binding model (there are two other groups of elements, the rare earths and the actinides, where a similar situation prevails; in their case the part-filled shells are the 4f- and 5f-shells). With such a starting point we can then go on to consider the effects of the overlap of these orbitals from atom to atom in the spirit of the linear combination of atomic orbitals (LCAO) approach we used in Chapter 2 for molecules. If the overlaps are small the spread in 'molecular' (crystal) orbital energy levels will be fairly small, just as the separation of the ψ_+ and ψ_- molecular orbitals for H_2 is small for small atomic orbital overlap. Thus following the considerations of crystal orbitals in Section 2.4 we will expect in

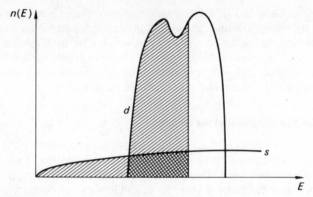

Figure 4.17   Overlapping s- and d-bands in a transition metal. This representation is only schematic; the shaded area (representing occupied states) in the d-band would be about 7 times that in the s-band for metallic iron.

a transition metal for each atomic d-orbital to have a narrow band of N crystal energy levels, if there are N atoms in the crystal. Since the 3d-like one-electron levels of transition metal atoms lie close (see p. 10) in energy to the 4s atomic levels that give rise to free-electron-like broad bands in metallic solids, we will expect our narrow band of d crystal levels to be overlapped by the broad band of conduction electron states derived from the 4s-levels. This concept of overlapping d- and s-bands shown schematically in Fig. 4.17 was introduced by Mott and was strikingly successful in providing a basis for discussions of transition metal properties. Now let us consider the character of these 'tight-binding' d-levels in a little more detail.

Suppose that we start with a one-dimensional chain of atoms that are well spaced compared with atomic dimensions (Fig. 4.18), and consider just one occupied state, described by the usual set of atomic quantum numbers n, l, m_l and m_s, on each atom. We appear to have N electrons all in the same state, in violation of the exclusion principle, but in reality this is not so for there is a fifth quantum number that takes precisely N different values: the position R_i; no two electrons have the same value of R_i. Individual states ϕ of the system, labelled by n, l, m_l, m_s and R_i, have large amplitude only in the vicinity of R_i. In quantum mechanical language we have a complete orthonormal (zero overlap between states of different R_i) set of states. It is always then possible to choose a different complete set of states; our interest is in the situation where electrons are not so well localized and can pass from atom to atom, so that the new set of eigenstates do not have R_i as a good quantum number. Even in the limit of large atomic spacing we can construct such eigenstates by taking combinations in which the probability of finding an electron is evenly distributed over all atoms:

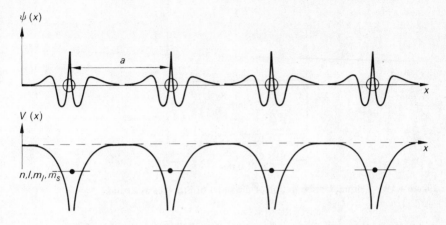

Figure 4.18   A chain of length $L = Na$ widely spaced atoms with one electron on each atom in the energy level n, l, m_l, m_s, represented schematically by the wavefunctions in the upper diagram and by the short horizontal line in each potential well.

$$\psi_k = \frac{1}{\sqrt{N}} \sum_i \alpha_{ik} \phi(n, l, m_l, m_s, R_i)$$

The quantity α_{ik} is a phase factor which will in general be different at each atomic site, and each combination wavefunction ψ_k will be characterized by a specific set of α_{ik}. The phase of a single atomic wavefunction ϕ considered in isolation is not a physically meaningful quantity, but phase *differences* are all important.

We must choose the phase factors in a systematic fashion so as to make sure that the delocalized ψ_k are a complete orthonormal set with just as many states, N, in it as the original set. The Bloch theorem suggests that the way to do this is:

$$\alpha_{ik} = \exp(ikR_i)$$

and the allowed values of k can be found by applying periodic boundary conditions to a chain of N atoms:

$$\exp(ikR_i) = \exp[ik(R_i + Na)]$$

therefore $k = m(2\pi/Na)$. m may take N values, and these may be chosen to run between $-N/2$ and $N/2$, so that k lies between $-\pi/a$ and $+\pi/a$, just as in the reduced zone scheme (§4.4).

The charge distribution of a ψ_k state is (remembering that so far there is no overlap)

$$| \psi_k(r) |^2 = \frac{1}{N} \sum_i | \phi(r - R_i) |^2$$

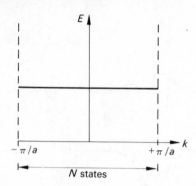

Figure 4.19   Formal representation of the states of Fig. 4.18 as a band.

which is just the same shape as that of the original ϕ states, but scaled down by a factor N on each site, and with an equal contribution from all N sites. The energies of all N ψ states are equal, just as all N ϕ states had equal energies, and of course they are equal to each other. This formal translation to a new set of states does allow us to draw an energy band of E versus k (Fig. 4.19); from a band point of view the group velocity is zero, meaning that electrons cannot flow along the chain because the atoms are so far apart.

If we now decrease the interatomic spacing to a point where there is significant overlap of atomic wavefunctions the electrons can move from site to site, because of the finite value of the wavefunction between atoms. Loss of localization means that atom-like wavefunctions are no longer eigenstates, and instead we are now forced to use the Bloch-like set of states. The different values of k express the phase differences between neighbouring atomic sites, and so affect the magnitude of the wavefunction between the atoms (Fig. 4.20) and therefore the charge distribution. The electrostatic energy of interaction between the electron state and the positive ions will therefore depend on k, and so will the kinetic energy; thus the band will now have a finite energy spread (Fig. 4.21).

Because the *number* of states is unchanged as the spacing alters it is clear that each tight-binding band has as many states as the original set of atoms. If the atomic states were s-orbitals, the s-band must contain one k-state per atom; a d-band contains five k-states per atom each holding two electrons, and so on.

The larger the overlap of atomic wavefunctions the greater the width of the energy band, also the position of the band on an energy diagram is determined by its occupancy, for if it is part-filled it must contain the Fermi level. In this way even all eleven electrons on each atom of sodium in sodium metal can be given a band description: the deepest levels have wavefunctions that overlap very little between atoms and form a narrow band that is full, further up the bands do have a finite width but are 'inert' because they are full. The outermost electron,

Re $[\psi(x)]$

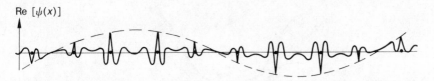

Figure 4.20   The real part of the wavefunction of a Bloch state formed by slight overlap of atomic 3s wavefunctions; $k = \pi/4a$. Comparison with Fig. 4.13 will indicate that while Re$[\psi]$ is modulated, $|\psi|^2$ (the charge density) is the same at each site. The dotted line represents the envelope e^{ikx}.

from a tight-binding point of view, is in a state with large overlap from site to site, so much so that the free-electron approximation is a better one. On the other hand, a transition metal d-band is of moderate breadth and, because of its tenfold degeneracy, contributes a very large density of states at the Fermi level, which is readily seen in the electronic specific heat (Table 3.2).

The combination wavefunction, which in three dimensions we can generalize to

$$\psi_k(r) = \sum_{R_i} \exp(ik . R_i)\phi(r - R_i)$$

is only a first approximation, and as the spacing decreases it does get modified, but, necessarily, the exact eigenstates always are Bloch states. A further complication arises from splitting of the d levels: in a free atom all ten d-levels have the same energy, but if the atom is in some lower symmetry environment, the different orientations of the charge densities of the various d-orbitals cause them to have different energies. This removal of degeneracy, known as crystal-field splitting, carries over into the metal, and there may be up to five d-sub-bands each doubly spin-degenerate. A final detail is that of mixing between the s-band and the d-band; again, in a free atom the s- and d-levels are orthogonal to each other and cannot mix, but the band states that derive from these atomic levels lose their orthogonality. The strongest mixing occurs between states from the two bands that have the same energy and the same value of k, that is, where

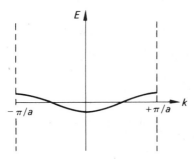

Figure 4.21   Energy versus k for a band of tight-binding states. The horizontal line represents the energy of the unperturbed atomic level.

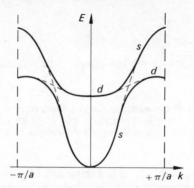

Figure 4.22   Hybridization between a nearly free-electron s-band and a tight-binding d-band.

the bands cross on an $E-k$ diagram (Fig. 4.22); in that region the k-states are hybrids, and the phenomenon is known as sd-hybridization. It is exactly analogous to the sp-hybridization that we encountered in bonding by C atoms in Section 2.4. (Sub-bands and hybridization effects can be seen in Fig. 4.25.)

Thus we can combine the conclusions we reached in Chapter 1 about the one-electron energies and wavefunctions (Fig. 1.6) of 3d- and 4s-states in transition metal atoms with the viewpoint outlined above to provide a useful picture of the spectrum of allowed electron states in solid transition metals. Fig. 4.23 shows schematically the broadening into bands of both 4s- and 3d atomic levels as the atoms are brought together into the crystal. At the observed interatomic distance the s-band must look much like the conduction

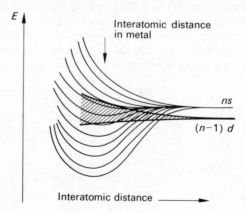

Figure 4.23   Schematic representation of broadening into bands of atomic s- and d-levels in transition metals.

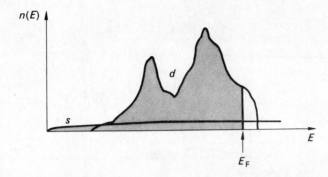

Figure 4.24   Density of states versus energy for a transition metal; d-band from tight-binding calculation, s-band from free-electron model; hybridization has been ignored. The value of E_F indicated is that for (paramagnetic) nickel.

band yielded by the 3s-levels of Na or the 4s-levels of Cu, but because of the smaller overlap of wavefunctions the 3d-band must be appreciably narrower. The density of states curves (in the absence of sd-hybridization) must consequently take a form somewhat like that shown in Fig. 4.24 where the structure reflects the presence of the sub-bands referred to above. The area under such curves up to E_F gives the numbers of electrons (n_d and n_s). For transition metals proper $0 < n_d < 10$ (and the Fermi energy lies in both bands), but even in Cu where all the physical properties indicate that the d-band is completely full the d- and s-bands must overlap and the colour of metallic Cu (due to a strong optical absorption in the blue end of the spectrum) suggests that the top of the d-band lies about 2 eV below the Fermi energy. If the bands did not move relative to one another or change their widths on passing from Cu to Ni, Co and Fe we would expect around 0·6, 1·5 and 2·4 empty 3d-band states per atom respectively in these metals, with around 0·6, 0·5 and 0·4 electrons per atom in the s-band. In fact relative movements of the bands that reflect those of the corresponding atomic levels yield s-electron counts of about 0·6, 0·7 and 1·0 per atom. However, as the atomic number decreases the influence of sd-hybridization becomes more important and it is increasingly difficult to make a meaningful distinction between s- and d-electrons. Fig. 4.25 shows: (a) $E-k$ relationships; and (b) the resultant density of states given by a theoretical calculation for Fe. It is now clear that deviations from free-electron character in soft X-ray bandwidths and specific heats like those indicated for transition metals in Tables 3.1 and 3.2 are to be expected on the tight-binding approach.

The high density of states of a transition metal makes for a large electronic contribution to the susceptibility, and is essential for the ferromagnetism of Fe, Co and Ni. It is important too for superconductivity in V, Nb, etc., and is a major factor in the large binding energies (and so the high melting points) of

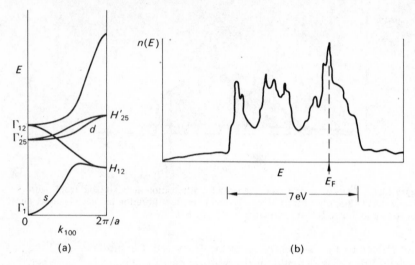

(a)   (b)

Figure 4.25   Energy bands for iron. (a) For k_{100} one of the branches from Γ'_{25} is doubly degenerate. Note the strong hybridization between the s-branch and one of the d-branches. (b) The total density of states; E_F marks position of Fermi level for paramagnetic Fe.

transition metals. It might also be expected to make for a high electrical conductivity (Chapter 3.5), but in fact the reverse is true and the transition metals have rather high resistivity. The reason is that the d-band electrons move sluggishly (the d-band has low group velocity) and contribute little to the current, indeed they hinder it by providing many states at the Fermi energy into which current-carrying s-electrons can be scattered.

Another situation to which the tight-binding approach seems applicable is the description of states of ionic crystals. In our earlier discussion of such materials (p. 25) we assumed that, in effect, the electron states of the filled ion core subshells (e.g. $2s^2 2p^6$ of Na^+ and $(3s^2 3p^6)$ of Cl^- were unchanged when the separate *ions* were brought together into a crystal. In fact, since the ion cores come into contact with one another some wavefunction overlap must take place and lead, in the spirit of the tight-binding approach, to low-lying full 2p-bands for the Na^+ ions, rather higher energy full 3p-bands for the Cl^- ions and empty high energy 3s-bands for the Na^+ ions. Since the first two of these sets of bands are completely full and the last completely empty in the Na^+Cl^- crystal no real change in our picture of the crystal is involved. However, if we go on to a substance like ZnS, application of the tight-binding methods shows that the broadening of the ionic levels 3p (S^{--}) and 4s (Zn^{++}) in the crystal is comparable with their separation in energy. The gap between the highest occupied state and the next available state is therefore much less in ZnS than in

NaCl, and in the former it is small enough for quanta of visible radiation to excite electrons across the gap and make it a photoconductor.

4.6   The calculation of band structures

In the preceding sections we have examined two extreme limits of possible approaches to the allowed states of electrons in crystals. In the nearly-free-electron approach the periodic part of the potential is treated as a minor modification to free-electron states; in the tight-binding approach the overlap of atomic wavefunctions from atom to atom is regarded as small enough to permit crystal wavefunctions to be generated in the spirit of the LCAO approach for molecules. It would seem likely that for many metals a correct approach would lie between these two extremes, using wavefunctions that close to a particular nucleus look like atomic wavefunctions but in between the nuclei are similar to the plane waves of free-electron theory, except for those values of k close to satisfying the diffraction condition.

Between 1930 and 1950 a great deal of ingenuity was exercized by theoretical solid state physicists in devising methods of constructing eigenstates of this character and calculating their energies. It was recognized that band structure calculations of this type were a prerequisite for discussions for example, of magnetic properties, transport properties, the possibility of superconductivity, etc. At first sight the problem seems an impossibly complicated one for there are strong interactions between the electrons and the positive ions and between the electrons themselves. (Although the periodic potential yields only diffraction, with no scattering at arbitrary k, a rough estimate of the first Fourier component of this potential in metallic Na yields about 5 eV, which is as large as the free-electron Fermi energy; also at their average spacing the conduction electrons have a Coulomb interaction of comparable magnitude.) An exact calculation would also have to include exchange between the electrons and to ensure that their states were properly orthogonal to those of the ion core.

When, however, indications came from experiment that for simple metals only very slight modifications of even free-electron models were required, there were grounds for hoping that justifications could be given for such drastic approximations, and these were soon found.

A full account of the methods of band structure calculations lies outside the scope of this book and the interested reader may consult a number of books and review articles†. It is of interest however to consider some of the concepts involved, although not necessarily in historical order.

†E.g. H. Jones *The Theory of Brillouin Zones* 1960 North-Holland; J. Callaway *Energy Band Theory* 1964 Academic Press; L. Pincherle, *Reports on Progress in Physics* 1960, **23**, 355; W. A. Harrison *Solid State Theory* 1970 McGraw-Hill.

Re $[\psi(x)]$

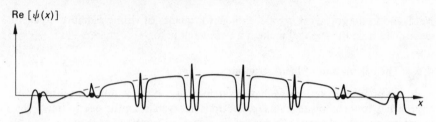

Figure 4.26   Real part of OPW wavefunction showing that $\psi \approx e^{ikx}$ for a large fraction of the crystal volume.

4.6.1   Orthogonalized plane waves

By making the conduction electron states orthogonal to those of the ion cores their wavefunctions can be forced to develop the 'atom-like' oscillations which must exist near the nuclei. The problem is that the high spatial frequencies required correspond to diffraction by large reciprocal lattice vectors. (In the analogous optical situation, in order to obtain all the detail in an image reconstructed from a hologram all orders of diffraction must be used. In our case the regular array of positive charges is the hologram that is generating the detail of the true wavefunction from plane waves.)

On the other hand, the experimental information on Fermi surfaces of simple metals that accumulated during the early 1960's gave a clear indication that the nearly-free-electron model is a good one for the simple (non-transition) metals. Within a simple orthogonalized plane wave (OPW) approach it is hard to see how this comes about, for the exact wavefunction has large additional kinetic energy near each atomic site (a large value of $\nabla^2 \psi$) and also a large and negative electrostatic energy in that region, and although these contributions tend to cancel, it is not obvious that the net effect should result in less than $0\cdot1\%$ distortion of the potassium Fermi surface from a sphere, particularly since in the (110) directions the Fermi surface is not far from the Brillouin zone boundary.

The resolution of this problem came with the realization that the smooth part of the true wavefunction (Fig. 4.26) differs little from a plane wave; instead of attempting to solve the Schrödinger equation for the exact wavefunction and the true potential an alternative and much weaker ion core potential, the pseudopotential, is used such that the solution found (a pseudowavefunction) closely resembles the smooth part of the true wavefunction; the cancellation between the two parts of the electron interaction, the high kinetic energy of an atom-like valence state and the low electrostatic energy of the deep ion core potential, are built into the problem at the beginning. The pseudopotential is further attenuated by dielectric screening, for the conduction electrons are mobile and respond to eliminate any long-range electric fields whether these are produced by local impurities or the conduction electrons themselves.

The pseudopotential approach was quick to explain the nearly-free-electron behaviour of the simple metals and provided accurate band structures for them, but the problem of the transition metals (and also Cu, Ag and Au) was a more formidable one because of the presence of both s- and d-electrons (as we saw in § 4.5 in Cu, Ag and Au the d-bands lie just below the Fermi energy) and the hybridization between them. Methods have now been developed which treat the s-electrons with a pseudopotential and simultaneously retain the localized character of the d-electrons.

Pseudopotential methods are now sufficiently finely developed that not only can accurate band structures be calculated for metals (and alloys) but also meaningful estimates can be made of the difference in energy of different crystallographic structures for an element. By including in the calculation information about the free-atom energy levels it is now proving possible, for the simple metals at least, to calculate the cohesive energy, the most favourable crystal structure, the equilibrium lattice spacing, and, of course, the band structure.

4.6.2   Cellular and related methods

An alternative approach to starting with plane waves and forcing them to develop atom-like oscillations at the nuclei, is to start by dividing the solid into cells centred on the nuclei and finding solutions (that connect smoothly from cell to cell) to the Schrödinger equation for the central potential in such a cell. Early methods of this sort used space-filling cells; but, recognizing that midway between a pair of atoms the potential must be flat for some distance, more recent methods have used 'muffin-tin' potentials (Fig. 4.27) that are constant outside spheres around the nuclei within which the potential is spherically symmetric. The wavefunctions take the form of plane waves between these spheres matched to atom-like functions on their boundaries, and are called

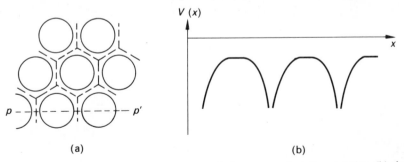

(a)   (b)

Figure 4.27   Muffin-tin potential for augmented plane wave calculation, (a) plan, (b) along the line $p-p'$.

augmented plane waves. Calculations by this method are very laborious and have become a practical possibility for a variety of solids only with the advent of high speed computers. An instructive approach is to express the problem in terms of a partial wave analysis of scattering of the plane waves by the ion cores. The scattering never becomes overwhelmingly strong because the cross-section fluctuates as the depth of the potential increases, bound states being drawn in to keep the cross section comparatively small. These bound states that keep the scattering weak are of course the ion core states which in the OPW approach justified the use of the pseudopotential.

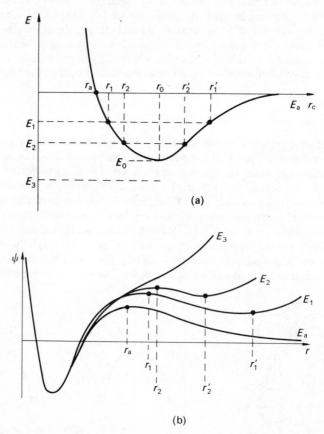

Figure 4.28   Wigner–Seitz method for wavefunction of lowest energy conduction band state in sodium. (a) Energies for which Schrödinger equation solved in atomic polyhedron and corresponding values of radius r_w satisfying $d\psi/dr = 0$. E_a is the energy for the isolated atom. (b) Wavefunctions for the energy values of (a).

4.6.3   The Wigner–Seitz method for the binding energy

In spite of the disadvantages of a cellular method with space-filling cells it is of both historical and physical interest to outline an early approach (1934) by Wigner and Seitz to the problem of the eigenstate and energy of the lowest energy conduction band state in a simple metal. They substituted for the polyhedral unit cell containing an atom at its centre (the Wigner–Seitz polyhedron of § 4.2) a sphere of equal volume of radius r_w and solved the Schrödinger equation for the spherically symmetric potential for various values of E subject to a boundary condition making $\partial\psi/\partial r$ vanish at r_w.

The resultant wavefunctions (Fig. 4.28b) and the corresponding values of r_w (Fig. 4.28a) are shown for Na. The resultant wavefunction for the crystal corresponding to $E = E_0$ is clearly (Fig. 4.29) a Bloch state with $k = 0$ and, subject to some very drastic assumptions about correlation, we can use this wavefunction in comparison with the atomic 3s-wavefunction of Na to see the origin of the metallic binding energy. In the free atom

$$\int_{\text{all space}} \psi_a^* \psi_a \, d\tau = 1;$$

in the metal there is on average one conduction electron per Wigner–Seitz sphere so

$$\int_0^{r_w} 4\pi r^2 \, \psi_m^* \, \psi_m \, dr = 1.$$

Hence the charge cloud corresponding to the shaded portion of Fig. 4.30 has been pushed into the more attractive potential V_m. The analogy with our interpretation of the origin of the binding energy of H_2 in terms of the occupation by both electrons of such a potentially attractive region is very close. (As in that problem much hinges on the electrons' having the good sense to keep out of each others way and minimize their repulsion. More formally we can say that the electron which occupies a given cell is surrounded by a correlation hole

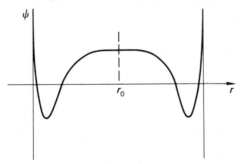

Figure 4.29   The lowest energy function for which a solution exists with connection to that in next polyhedron.

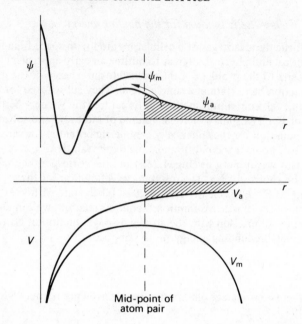

Figure 4.30   Schematic illustration of the origin of binding in Na metal. The charge density corresponding to the shaded part of the atomic ψ has been pushed back into the atomic polyhedron which can be seen to provide a more attractive potential for it than V_a.

of radius r_w.) It is important to realize that the state we have been considering is that at the bottom of the band ($k = 0$), and the binding energy of sodium per atom (hence per valence electron) will be less than $E_a - E_0$ by the average band energy

$$\frac{\int_0^\infty n(E)\,f(E)E\,dE}{\int_0^\infty n(E)\,f(E)\,dE} = \overline{E}$$

which for free electrons ($n(E) = $ constant $\times E^{1/2}$) will be $\frac{3}{5}\,E_F$ at 0 K.

This simple approach gives reasonable values of both the cohesive energy and the equilibrium atomic volume for Na and K.

4.7   Collective electron models for alloys

In view of the rather surprising success of free-electron and nearly-free-electron models for simple pure metals, it seemed not unreasonable to apply similar concepts to simple binary solid solution alloys. That is to say the situation should not be drastically altered if some of the electrons in the electron gas were contributed by atoms of type A and some by atoms of type B. If these atoms are distributed at random over the points of a simple lattice (possessing perhaps the symmetry of pure A or pure B or perhaps that of some other simple structure), there will be a loss of perfect translational symmetry, but the mobile electrons can be expected to screen out the grosser effects, and make each atomic polyhedron approximately electrically neutral. That the addition of 10% of Au or Zn atoms to Cu does not change its room temperature electrical resistivity by much more than a factor of two is confirmatory evidence that the effects are not too drastic.

If this approach is justified it means that we can change the average band occupation in a metal by alloying it with some metal of different valency while retaining the reciprocal lattice, the Brillouin zone, the constant energy contours, the band gaps, etc., of the host metal. To assume that these quantities are quite unchanged yields what has been called the *rigid-band model* of an alloy. To assume that we have a collective band to which both elements contribute electrons, but the parameters of which (as well as the occupation) are a function of alloy concentration is a more reasonable (*soft-band*) assumption. The solid solutions of elements like Zn, Ga, Ge, Al and Sn in Cu provide useful testing grounds for such approaches since no empty d-levels are present. Recently there have been direct measurements of changes in Fermi surface dimensions with alloy concentration that are in reasonable agreement with rigid-band predictions based on contributions of 2, 3, 4, 3 and 4 conduction electrons per atom of these solutes respectively.

The metallurgical character of these alloy systems was in fact an important prop to early arguments (before measurements of the Fermi surface were possible) for nearly-free-electron approximations and rigid-band models, because Hume-Rothery had observed that the solid solubility limits, the compositions of intermediate phases of given character, and other features pointed to the average number of conduction electrons per atom as a controlling parameter. Perhaps the most surprising feature of these results was that the limit of solid solubility of polyvalent (non-transition) metals in f.c.c. Cu, Ag or Au, after which a b.c.c. phase appeared, seemed to correspond (if other factors were favorable to alloy formation) to values of electron/atom ratio for which the Fermi surface of a free-electron gas would just touch the faces of the f.c.c. Brillouin zone. Since the mean energy per electron is slightly less (on a nearly-free-electron model) than that for a free-electron gas when the Fermi surface has almost touched a Brillouin zone face, and is slightly more shortly after (in the sense of increasing numbers of electrons) the Fermi surface has touched the Brillouin zone (Fig. 4.10), it seemed that this contribution to the energy was just tipping the balance in the control of the choice of crystal structure. Although this approach† appears

---

†It is interesting to note that H. Jones in his early theoretical work on this problem assumed a distortion of the Fermi surface of pure Cu very close to that experimentally identified 20 years later.

inconsistent in that it utilizes the energy changes by Bragg reflection on the one hand, but uses the free-electron model to determine the electron concentration at which Bragg reflection first occurs, in practice it works quite well, and more recently has been given sound theoretical justification by the careful calculation of band structures (see § 4.5).

The small electrical resistivities of these alloys represent a scattering cross section presented to the conduction electrons by the solute atom that is much less than atomic dimensions, for example the cross section for a zinc atom in copper is about $0\cdot2$ Å^2, whereas the atomic radius is about $1\cdot4$ Å and might be expected to represent a target of area 6 Å^2. However, as far as the conduction electrons are concerned Cu and Zn atoms differ only by the size of their positive charge; the complement of core electrons (filled shells up to and including 3d) is identical. The conduction electrons are attracted into the low potential energy region surrounding a solute Zn atom (Fig. 4.31), but they must ensure that the

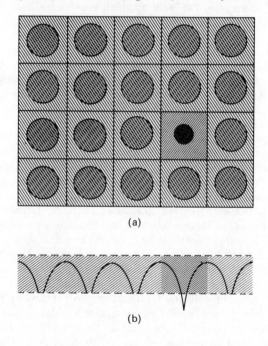

(a)

(b)

Figure 4.31   Modified conduction electron density around a zinc atom in the copper lattice (schematic).

exclusion principle is not violated, that is, their wavefunctions must be orthogonal to the ion core electrons on the Zn atom. However, the conduction electrons in pure Cu certainly have wavefunctions that are already orthogonal to the Cu ion cores, so very little adjustment is required in the conduction electron wavefunction to become orthogonal to a similar set of core electrons on the solute atom (the radii of the core wavefunctions on a Zn atom will be slightly less than those on Cu because of the increase in nuclear charge).

In a metal the mobile conduction electrons always respond quickly to any excess charge and adjust their densities so as to screen it out. If the core states of solute and solvent atoms are similar, little modification of the conduction electron wavefunction is required in order to do this, and the accompanying scattering is also kept very small. Conversely, large resistivities are found in alloys of elements that are chemically dissimilar (Fig. 4.32), and an approach quite different from the rigid-band model is then required.

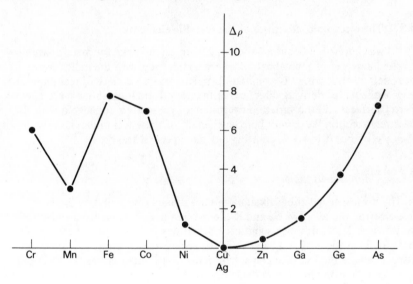

Figure 4.32   Increase $\Delta\rho$ in resistivity (in μ ohm cm) of Cu at room temperature produced by 1 atom % of solute elements in the same row of the periodic table, and of Ag in the next row.

Anything like a nearly-free-electron band structure must clearly fail for the alloys of the transition elements, as it does for those elements themselves. It is worth noting, however, that a *collective* band structure does seem to be indicated by many effects, and there are a remarkably large number of binary transition metal–transition metal alloy systems for which the average d-band occupation is a useful parameter against which to plot the magnetic susceptibility, the electronic specific heat or even the superconducting transition temperature. (The famous plot of average saturation magnetic moment against average atomic number – the Slater–Pauling curve – for ferromagnetic binary alloys of neighbouring elements along the series of elements Mn–Fe–Co–Ni–Cu is the oldest established example. A more detailed discussion of these data will be found in *The Magnetic Properties of Solids* by J. Crangle, in this series.)

Solid solution alloys of transition metals or rare earths in simple metals present a much more complicated situation, and one that has only been clarified quite recently. One must expect that a Cu—1% Fe alloy will have d-like solute atom levels near the Fermi energy, and that while no d—d overlap between solute atoms is likely, these levels will be broadened by mixing with the conduction electron states (in analogy with the sd-hybridization in the band structures of pure transition metals (§ 4.5)). The resultant 'resonant' levels give rise to strong conduction electron scattering (Fig. 4.32) and to rather complicated magnetic behaviour.

4.8   The electronic structures of the metallic elements

As was implied in Section 4.6 energy band calculations for metals have now reached a degree of sophistication where comparison with the indications provided by experiment is meaningful. Furthermore a variety of techniques is now available for deriving direct experimental information about the shapes of Fermi surfaces and the variation over them of parameters connected with the $E-k$ relationship. We cannot here give details of these techniques and the results they yield† but it is worth pointing out some general trends.

4.8.1   Monovalent metals

The body-centred cubic alkali metals are remarkably well described by free-electron models. For Na and K the maximum deviations from free-electron k_F values is $0·1\%$, although significant deviations from spherical Fermi surfaces exist for Li at the beginning and Cs at the end. Even for the [110] direction in k-space, where the free-electron sphere most closely approaches the Brillouin zone boundary, the maximum deviations of k_F from the free-electron value are only about 4% for Li and 8% for Cs as compared with the 12% that would be needed to establish contact with the zone face.

The face-centred cubic metals Cu, Ag, and Au with configurations $ns^2 \, np^6 \, nd^{10} \, (n+1)$ s have Fermi surfaces that show appreciable distortions from sphericity and in all of them the Fermi surface makes contact (the 'necks') with (111) faces of the Brillouin zone. The energy gaps in $E(k)$ across these faces are much larger (~ 4 eV) than those in the alkali metals and this is certainly due to the effects of hybridization between $(n+1)$ s-band states and just-filled nd-band

---

†For an excellent survey of the principles see A. B. Pippard, *Reports on Progress in Physics* 1960 Vol 23 and for a recent comparison of experimental data and theoretical results see A. Cracknell, *The Fermi Surfaces of Metals*, 1973, Taylor and Francis.

states. Although the d-bands are full they lie at only 2–3 eV below the Fermi level and play a dominant role in many of the properties of these metals.

4.8.2   Divalent metals

Zn and Cd (which follow Cu and Ag in the periodic table) show more resemblance to Mg which follows Na than do Cu, Ag, and Au to Li, Na and K because the full d-shells in the former lie at fairly low energy and play little role in the properties. At two electrons per atom a free-electron sphere must intersect the faces of the first Brillouin zone (§ 4.3). Since Zn, Cd, and Mg are all close-packed hexagonal with two atoms per unit cell the basic Brillouin zone holds only one electron per atom (two electrons per unit cell of the crystal) and the free-electron sphere must occupy parts of at least three Brillouin zones, occupied states in the third and perhaps higher zones being equal in number to empty states in the second and perhaps first zones. For Mg the modifications from a free-electron sphere are not very great, but for both Zn and Cd the larger energy gaps across the Brillouin zone faces and the departures of the crystal structures from ideal close-packing (c/a = 1·633) produce significant deviations, particularly in Cd where certain portions of the Fermi surface that would be expected on a nearly-free-electron model do not appear at the observed 1·886 c/a value. The resultant areas of the Fermi surfaces are only about 40% of the free-electron values for both metals.

The alkaline earth metals have a number of allotropic modifications and less is known about the details of their electronic structures, but it is of particular interest that f.c.c. Ca and Sr have much smaller pockets of holes in the first zone and of electrons in the second than a nearly-free-electron model would predict, and the marked increases in electrical resistance of these metals with pressure suggest that the overlap becomes even smaller as the volume is reduced.

4.8.3   Polyvalent metals

Quite a lot of information is now available about the details of the electronic structures of metals with more than two electrons per atom in bands derived from atomic s- and p-states. Some of these (like f.c.c. Al and Pb) have fairly simple crystal structures and the reasonably close approximation of their electronic structures to estimates based on a nearly-free-electron approach can be shown without too much difficulty. (For details see the review by Cracknell quoted on the previous page.)

Others have more complicated structures ranging from the slight tetragonal distortion of a f.c.c. structure in In to the complex orthorhombic structure of Ga with eight atoms per unit cell, and some progress has been made in justifying the existence of these structures in terms of the details of the electronic structures.

The group V semi-metals As, Sb and Bi are of particular interest, since the electronic structure of Bi was discussed early in the history of Brillouin zone theory by H. Jones in terms of a nearly-free-electron viewpoint. He identified a zone (a Jones zone in the sense of § 4.4) which, when full, would accommodate five electrons per atom, and the surfaces of which lay everywhere close to a free-electron sphere containing five electrons per atom. Early estimates by Jones of concentrations of holes inside this zone and electrons outside of 10^{-3} to 10^{-4} per atom have been modified to about 10^{-5} per atom in more recent analyses of a variety of experiments. The exact location and shape in k-space of the pockets of electrons and holes has been a subject of much debate and we shall not give details. The Fermi surfaces of As and Sb, although less studied, are apparently similar to that of Bi, but the ordering in energy of various points in the zone is not identical. A striking consequence of this different ordering is that solid solution alloys in the As–Bi system show semiconducting behaviour for a certain range of compositions; clearly, for these alloys all states within the Jones zone have lower energy than all states outside it, although that is not the case for either of the pure component metals.

4.8.4   Transition metals

Nearly-free-electron models are not (as we might expect from our earlier discussions in § 4.5) a helpful guide to metals with large amounts of d-character at the Fermi surface. Band structure calculations (especially by APW methods) have, however, been refined to a point where they provide useful guidance to the interpretation of experimental data related to the Fermi surfaces, and the implications of these for magnetic and superconducting properties are of great interest. The type of band structure model useful for qualitative discussions of properties was shown in Fig. 4.25 and discussed on p. 89.

4.8.5   Germanium, silicon and carbon

The face-centred cubic lattice of Ge, Si and C (diamond) has two atoms in the primitive unit cell of the crystal, and provides a tetrahedral environment for each atom. (Referring our coordinates to the cubic cell we can see that an atom at ¼, ¼, ¼ is tetrahedrally coordinated by the atoms at 0, 0, 0; ½ 0 ½; 0 ½ ½; and ½ ½ 0.) In consequence the four electrons *per atom* that provide a full valence band (or in covalent bonding terminology occupy the bonding (sp^3) hybrid orbitals) can just fill four Brillouin zones and be separated by a gap – 0·67 eV for Ge, 1·14 eV for Si and 5·33 eV for diamond – from states in the next Brillouin zone which is empty.

Detailed band structure calculations for Ge are not violently different in their E–k relationships from what might be expected for a nearly-free-electron treatment of the diamond structure, with energy gaps of a few electron volts across the zonal faces and a net gap between the highest (full) fourth zone state and lowest (empty) fifth zone state. In Si all the energy gaps are larger and calculations show an interesting consequence of the sp-hybridization for the position of the lowest empty state in the fifth zone. In Ge this was (as might be expected from Fig. A.4.1b, p. 104) at the centre of the (111) faces of the Brillouin zone. In Si the energy at that point has been pushed up and the lowest energy state in the fifth zone is close to *but not exactly at* the centre of the square (200) faces of the zone.

At this point we should mention a feature of band structure calculations that becomes important when finer details are relevant. At the top of the full band in Si and Ge a level of pure p symmetry exists. The orbital angular momentum of this $l = 1$ level couples to the spin of an electron occupying it and, as in an atom with a single electron in a p-state, there exist two levels of slightly different energy with $j = \frac{3}{2}$ and $j = \frac{1}{2}$, the former fourfold degenerate ($m_j = +\frac{3}{2}, +\frac{1}{2}, -\frac{1}{2}, -\frac{3}{2}$,) and the latter twofold degenerate. Notice that these six states are one-electron states in a self-consistent field sense and are labelled with lower-case j. As in free atoms spin–orbit coupling effects increase with atomic number and the $p_{3/2}, p_{1/2}$ splitting at $k = 0$ is 0·044 eV in silicon and 0·29 eV in germanium.

In the diamond and graphite forms of carbon all connection with a nearly-free-electron model is lost, although for the former a relationship with the band structures of Ge and Si can be brought out by increasing the zone face energy gaps to large values. In graphite recent calculations and Fermi surface studies show a small overlap between a full (bonding) band and an empty (antibonding) band (as in As, Sb, and Bi) with a very small concentration of electrons and holes at 0 K and an electronic specific heat as small as 13×10^{-6} J mol^{-1} K^{-2}, which is to be compared with the 'metallic' values shown in Table 3.2 (p. 55). We saw in Section 2.4 that the highest energy bonding levels and the lowest energy antibonding levels in graphite (as in benzene) would be π bonds, for which we had seen that the bonding/antibonding energy separation is smaller than for σ bonds of the type that yield the bonding and antibonding bands of diamond. Thus both the zone language and the bond language make it reasonable that these two allotropes of carbon should have striking differences in conductivity, graphite being a semi-metal and diamond an insulator (cf. Fig. 2.14).

A.4.1   Appendix: Zone structures for the common metallic structures

In Section 4.3 we indicated the alternative descriptions of the body-centred cubic structure in terms of a primitive unit cell containing one atom, and that in terms of the cubic unit cell that contains two. The first (or basic) Brillouin zone contains N states where N is the number of unit cells in the crystal and it must be emphasized that this is the number of *primitive* unit cells. There are of course structures that (unlike the b.c.c. and f.c.c. structures of simple metals) contain more than one atom in even the primitive unit cell, so that the number of electrons that can be accommodated in the basic Brillouin zone is less than two *per atom*.

We show in Fig. A.4.1 the first Brillouin zones for the body-centred cubic, face-centred cubic, and hexagonal latttices. The faces are labelled with the Miller indices (referred to the conventional unit cell) of the crystal planes responsible for the corresponding diffraction.

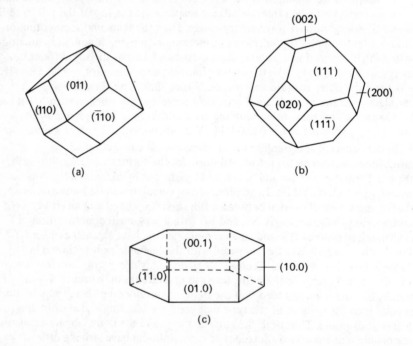

Figure A.4.1   The first Brillouin zones for (a) the body-centred cubic structure, (b) the face-centred cubic structure, (c) the close-packed hexagonal structure.

Fig. A.4.2 shows volumes in k-space bounded by diffraction planes for the close-packed hexagonal structure which contains two atoms per primitive unit cell of the lattice. These can all be used as Jones zones. (The true Brillouin zone, Fig. A.4.1c, has the property that the energy gap vanishes on the (001) type faces due to the structure factor.) In Fig. A.4.2, the zones (a) and (b) are twice the volume of the basic Brillouin zone and can therefore hold two electrons per

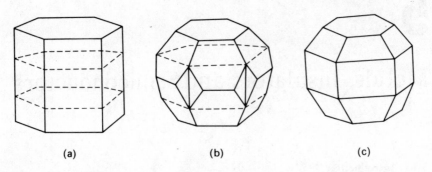

Figure A.4.2   Jones zones for the close-packed hexagonal structure (the faces of the re-entrant portion in (b) constitute extensions of faces of the type (01.0)). In (a) and (b) the Brillouin zone of Fig. A.4.1 (c) is shown dotted.

atom, since they hold four electrons per unit cell but there are two atoms in the primitive unit cell of the hexagonal close-packed structure. Only (b) however is the second Brillouin zone in terms of the definition on p. 70. The zone (c) is an example of a Jones zone bounded on all faces by energy gaps but holding less than two electrons per atom.

5

Metals, Insulators and Semiconductors

5.1 Introduction

The electrical resistivity of the solid elements provides a sharp distinction between metals and non-metals (and it is one that would be even sharper if the comparisons were made at low temperatures); pure metals rarely have resistivities exceeding $100\ \mu\Omega$ cm at room temperature and it is unusual to find a disordered metallic alloy that has resistivity greater than $300\ \mu\Omega$ cm. On the other hand, semiconducting elements have resistivities that are orders of magnitude greater, and diamond has an ideal resistivity at least 10^{20} times that of a metal. An element can be on one side of the metal/semiconductor boundary in one allotropic form, and on the other side in another. For example, grey tin is a semiconductor and, where it is stable (just below room temperature) has a rather high resistivity, but the metallic allotrope, white tin, is a good metal. This sensitivity to structure extends also to differences between solid and liquid elements; for example, on melting the semiconductor Si the resistivity drops by a factor of 10^2, and it becomes a reasonable metal.

The sensitive dependence of electrical properties on *crystal* structure suggests that for some forms of *band* structure the electrons are capable of carrying current, and for others they are not. We must therefore first look in detail at the way in which band structure can affect the conductivity.

5.2 Full and empty bands

In Section 3.5 we saw that a k-space description of the conduction of electrical current involved the displacement of the electron distribution by an applied electric field. Consider what happens if the k-space displacements are (unrealistically) large (Fig. 5.1), and bring occupied states up to the band gap. The electrons have far too little energy to jump the gap, and instead are forced to remain in the lower band; we can draw this either in the repeated (Fig. 5.1b) or reduced (Fig. 5.1c) zone schemes. What is happening in real space is that the field accelerates an electron as it would a free particle until the diffracted wave starts to build up; there is more and more reflection as the band gap is approached until at state X (Fig. 4.15) there is just a standing wave (which

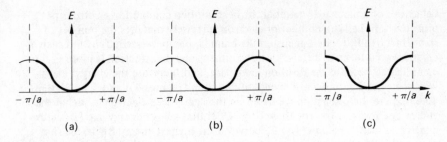

(a)   (b)   (c)

Figure 5.1   (a) The equilibrium distribution of electrons in a part-filled metallic band; (b) given a large displacement in k-space by an electric field; (c) represented in the reduced zone scheme.

cannot carry current). State X' is the same state as X, and, under the influence of the electric field, the electron distribution continues to move steadily through k-space.

If, however, the band happened to be full (Fig. 5.2), there could be no displacement of the occupied states. In the repeated zone picture there appear to be states available between X' and Z' but these are in fact the occupied states between X and Z and in the reduced zone scheme it can be seen that no real change is possible. A full band cannot carry current. We can describe this another way: any band must, by symmetry, contain a number of states with velocities to the left equal to that with velocities to the right; if all the states are occupied the current is certainly zero, and cannot change from zero without the introduction of the large amount of energy required to excite electrons up to the next band.

Now the exact relation between the number of states in a band and the number of unit cells in the sample ($\S$ 4.2) comes into play in governing the

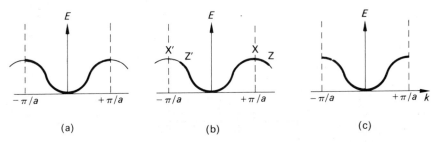

(a)   (b)   (c)

Figure 5.2   As in Fig. 5.1, but now with a full band. The displaced distribution (b) is identical to the equilibrium distribution (a), as can be seen by plotting it in the reduced zone scheme (c).

behaviour of a particular material: if there are two conduction electrons per primitive unit cell the number of electrons precisely matches the number of states in a band. In one dimension the band is then necessarily full, and such a crystal would be an insulator. In three dimensions the situation is more complicated because the position in k-space, and therefore the energy, at which the band gaps occur depend on direction (Figures 4.4 and 4.7). We can take up here the discussion we based on the nearly-free-electron model at the end of Section 4.2. We saw there (Fig. 4.12) that a large energy gap (Δ) across zone faces leads to a *band gap* E_g between the highest energy E_R in the first zone and the lowest energy state E_Q in the second zone. With exactly two conduction electrons per unit cell (and large Δ) the whole band of states lying in the first zone will be full and those in the second zone will be empty (Fig. 5.3). The crystal will necessarily be an insulator at 0 K. In semiconductor terminology the full band is called the *valence band*, and the empty band the *conduction band*. How high a temperature must be achieved before the Fermi–Dirac statistics gives a finite probability of occupation of states above the gap depends on the size of the gap and this depends on the atomic character of the crystal. With small values of Δ, however, E_Q will be less than E_R (Fig. 4.10) and metallic character follows. Thus for divalent elements no general prediction of the character of the solid is possible. Either insulating or metallic behaviour is compatible with general theories; which is found for a particular element could depend on the details of the atomic energy levels, the crystal structure and the interatomic distance.

We have already seen that the more easily an element ionizes and loses its outer electrons the smaller we expect the band gaps to be; so the group II elements Be, Mg, Zn and Cd do form good metals despite having two electrons per atom. For reasons connected with their proximity to the transition elements

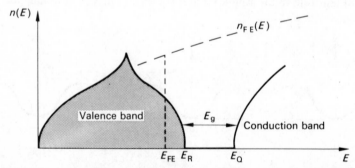

Figure 5.3 The density of states when the energy gaps across Brillouin zone faces are large (cf. Fig. 4.12). With two (or an even integer) electrons per unit cell the valence band is fully occupied, and is separated by a *band gap* E_g from the empty conduction band. E_{fe} represents the energy to which a free electron band would be filled for the same number of electrons.

Ca, Sr and Ba are not very good metals, and there is evidence that the band structure is sufficiently close to that of an insulator for them actually to become so under pressure. The elements of group IV provide another illustration of these ideas: C, Si and Ge are all insulators at absolute zero, but the band gap decreases steadily down the series; Sn is a borderline case with one allotropic form, gray Sn, having a small band gap, and the other, white Sn, being metallic. The last element in the group, Pb, is always metallic. The frontier between insulator and metal coincides with a change in chemistry; C, Si and Ge characteristically form covalent compounds in which electrons are shared, charged C ions are unknown in stable chemical combination; Sn and Pb do ionize in forming their salts, but there is still a residual tendency to share electrons (for example the stannates and plumbates). We saw in § 2.6 that the bonding in these group IV elements becomes weaker as we go down the series, metallic bonding being less effective than covalent bonding, and in Table 3.1 that the experimental values of the occupied band widths were in fact larger for Si and Ge than those expected on a free-electron model. (This is to be expected because $n(E)$ lies below the free-electron value for the upper portion of a full band, as shown in Fig. 5.3, so correspondingly the band must be occupied to a greater energy.) The increasingly metallic character as we go down the series does not greatly modify the occupied bandwidth, but it yields a very important reduction in the energy gap Δ across the zone faces (corresponding to $(E_Q - E_P)$ in Fig. 4.12) and hence to a reduction and eventually a disappearance of the band gap $E_Q - E_R$).

It is easily seen that large energy gaps, and hence the possibility of insulating behaviour, must also follow if all outer electrons in a crystal are in closed-shell states that overlap only slightly from atom to atom and are most appropriately described by tight-binding wavefunctions. An extreme example is provided by the monatomic close-packed crystals of the rare gases where a narrow full band (derived from the np^6) electron shell of the atom is well separated from the empty bands that are formed by overlap of $(n + 1)$ s-states.

In simple ionic crystals the use of a Bloch function and energy band description is still possible but the closed shells must yield a similar situation to that of the inert gases, although in Na^+Cl^- the highest energy full band will be derived from (p^6) full shells of Cl^- while the empty band wavefunctions will be produced by overlap of 3s atomic functions of Na.

At this point we should perhaps draw attention to the fact that also in *liquid* rare gases the band of levels derived from the full p^6-shell will not overlap the band derived from the empty s-levels, and hence these liquids will be insulators. The significance of the fact that loss of lattice periodicity does not necessarily lead to metallic conductivity will be discussed in more detail in Section 6.2.

Summary

In this section we have seen that the sharp distinction in properties between metals and insulators arises from distinct differences in the occupation of bands, in the latter all bands are either completely full or completely empty, and in crystalline solids there is a one-to-one correspondence between a full band of states and a full Brillouin zone. Semiconductors form a class distinguished qualitatively from the insulators by the narrowness of the forbidden range of energies separating allowed and full bands from allowed but empty bands. It is important to notice that for simple translational lattices (one atom per primitive unit cell) an odd number of electrons per atom means that metallic conductivity must exist since some Brillouin zone or zones must be partly filled. With an even number of electrons per atom such a crystalline solid may or may not be a metal or insulator at 0 K depending on whether the energy gaps across zone faces are small (e.g. Mg) or large. No element comes into the latter category (Ge and Si have two atoms per unit cell, Se and Te have three) but as pointed out in Section 4.7, Ca has large enough gaps to have only a small overlap of bands and may become a semiconductor at high pressures.

When there are two atoms per unit cell of the crystal structure an odd number of electrons per atom can lead to a situation of bands all full or all empty. Thus As, Sb and Bi with five electrons per atom are almost of this type being semi-metals with equal (small) numbers of electrons in one zone and holes in another which just overlaps it in energy, and some of their alloys with one another are semiconductors (see §4.8).

5.3   Semiconductors: electrons and holes

At absolute zero there is a sharp distinction between a metal and an insulator; if each were ideally pure the former would have zero and the latter infinite resistivity. At room temperature, many materials that we would classify as insulators do have a finite electronic conductivity because a small proportion of the electrons are thermally excited across the gap between full and empty bands leaving a part-filled band (the valence band) behind them and part occupying a new band (the conduction band). Because both these bands are incomplete both can carry current.

The Fermi—Dirac distribution function (Equ. 3.7, § 3.4) must describe the probability of occupation of states; because band gaps are rather large (of order 1 eV) compared with room temperature thermal energies (about 1/40 eV), the number of electrons excited is minute compared with the total, and it is more convenient to describe the valence band in terms of the distribution $f_h(E)$ of empty states (known as 'holes')

$$f_h(E) = (1 - f(E)) = 1 - \frac{1}{\exp[(E - \mu)/k_B T] + 1}$$

$$= \frac{1}{\exp[-(E - \mu)/k_B T] + 1} .$$

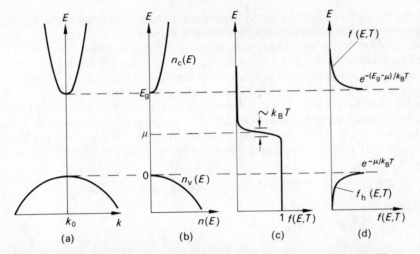

Figure 5.4   The band structure (a) and density of states (b) of a semiconductor. The Fermi–Dirac distribution function is shown in (c); note that $k_B T \ll E_g$. The distribution functions for electrons and holes are shown on a greatly expanded scale in (d). The densities of electrons and holes are given by the convolution of (b) with (d).

The chemical potential μ is that energy at which a state has probability ½ of being occupied; since the highest energy states in the valence band have only a small chance of being empty, and the lowest energy states in the conduction band a small chance of being filled, the chemical potential must lie within the band gap (Fig. 5.4). We can now calculate the total number of electrons in the conduction band in terms of the density of available states $n_c(E)$ as

$$n = 2 \int_{E_g}^{\infty} n_c(E) f(E, T) dE \tag{5.1}$$

where the energy has been measured from the top of the valence band. Similarly the number of holes in the valence band (density of states $n_v(E)$) can be calculated as

$$p = 2 \int_{-\infty}^{0} n_v(E) f_h(E, T) dE \tag{5.2}$$

Because the band gap is large compared with $k_B T$ the distribution functions can safely be simplified:

$$f(E, T) \simeq \exp[-(E - \mu)/k_B T]; \quad f_h(E, T) \simeq \exp[(E - \mu)/k_B T]$$

Also, it is useful to have an explicit form for the densities of states. Here we are interested in E as a function of k in the neighbourhood of a band gap; perturbation theory shows $E(k)$ to be parabolic in form, an upward parabola for the conduction band, and a downward one for the valence band. If we call the k-space position of the band gap k_0 (we are assuming that the top of the valence band and the bottom of the conduction band are at the same point in k-space, as in Fig. 5.4, this is known as a direct gap situation; for many semiconductors this is not so, and the gap is indirect (see p. 103, § 4.8); however, the essential physics is well illustrated by the more simple direct gap case,) we can write the two energy bands in a form that parallels the free-electron expression $E = \hbar^2 k^2/2m$:

$$E = E_g + \hbar^2 (k - k_0)^2/2m_c^* \qquad \text{for the conduction band}$$

and

$$E = -\hbar^2 (k - k_0)^2/2m_v^* \qquad \text{for the valence band}$$

In place of the free-electron mass m the band structure has been parameterized in terms of an effective mass m^*, whose physical significance will become apparent when we consider the dynamics of electrons and holes. A large m^* describes a nearly flat $E(k)$; in principle m^* can be obtained from a band structure calculation. The density of k-states could now be found in exactly the same manner as the free-electron case (§ 3.1), by considering spherical shells centred on k_0, but the parallel allows us to write the result immediately:

$$n_c(E) = (1/4\pi^2)(2m_c^*/\hbar^2)^{3/2}(E - E_g)^{1/2} \qquad (5.3a)$$

$$n_v(E) = (1/4\pi^2)(2m_v^*/\hbar^2)^{3/2}(-E)^{1/2} \qquad (5.3b)$$

The integrals are now

$$n = 2 \int_{E_g}^{\infty} \frac{1}{4\pi^2} \left(\frac{2m_c^*}{\hbar^2} \right)^{3/2} (E - E_g)^{1/2} \exp[-(E - \mu)/k_B T] \, dE$$

$$p = 2 \int_{-\infty}^{0} \frac{1}{4\pi^2} \left(\frac{2m_v^*}{\hbar^2} \right)^{3/2} (-E)^{1/2} \exp[(E - \mu)/k_B T] \, dE$$

which can be put in a standard form by a change of variable

$$n = \frac{1}{2\pi^2} \left(\frac{2m_c^*}{\hbar^2} \right)^{3/2} (k_B T)^{3/2} \exp[-(E_g - \mu)/k_B T] \int_{0}^{\infty} z^{1/2} e^{-z} \, dz;$$

$$z = (E - E_g)/k_B T$$

$$p = \frac{1}{2\pi^2}\left(\frac{2m_h^*}{\hbar^2}\right)^{3/2}(k_BT)^{3/2}\exp(-\mu/k_BT)\int_0^\infty z^{1/2}e^{-z}dz; \qquad z = -E/k_BT$$

The definite integral has the value $\sqrt{\pi}/2$, so that finally:

$$n = \frac{1}{4}\left(\frac{2m_c^*}{\pi\hbar^2}\right)^{3/2}(k_BT)^{3/2}\exp[-(E_g - \mu)/k_BT]$$

$$p = \frac{1}{4}\left(\frac{2m_h^*}{\pi\hbar^2}\right)^{3/2}(k_BT)^{3/2}\exp(-\mu/k_BT)$$

(5.4)

Since we started at $T = 0$ with a full valence band and an empty conduction band the number of holes must equal the number of electrons, which allows us to calculate μ:

$$\mu = \tfrac{1}{2}E_g + \tfrac{3}{2}k_BT \ln(m_h^*/m_c^*)$$

At low temperatures ($k_BT \ll E_g$, which is well satisfied at room temperature for nearly all semiconductors) the chemical potential is therefore in the centre of the gap and the exponential factors of Equ. 5.4 become $\exp(-E_g/2k_BT)$.

The number of carriers varies exponentially with temperature, being controlled by the factor $\exp(-E_g/2k_BT)$; the scattering rate for these carriers will be temperature-dependent, but only with some power law, consequently the conductivity itself will also have an exponential dependence, and this is what is experimentally observed in a pure semiconductor (Fig. 5.5). A semiconductor has a resistivity quite unlike a metal, for it is large and decreases rapidly with temperature; the temperature dependence gives one method of estimating the band gap. The parameters appropriate to a number of the more common semiconductors are given in Table 5.1; the density of carriers at room temperature in each of these materials has also been calculated, and it can be seen that it is many orders of magnitude lower than that for a metal. In the larger band gap materials, such as Si and GaAs, impurity effects dominate at room temperature and the observed carrier concentrations are considerably higher. A semiconductor in which the temperature is high enough and the

Table 5.1 Band gaps and densities of carriers at room temperatures for some semiconductors.

Semiconductor	Ge	Si	GaAs	InSb
E_g(eV)	0·7	1·1	1·4	0·18
Calculated intrinsic room temperature carrier concentration (cm^{-3})	10^{13}	10^9	10^7	10^{17}

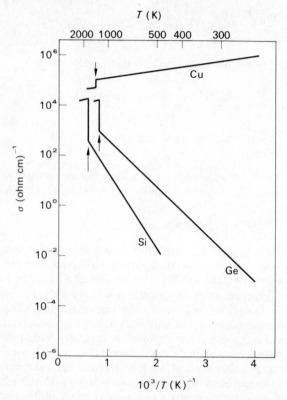

Figure 5.5   The conductivity of pure Si and Ge as a function of temperature. At lower temperatures than those shown here the conductivity is controlled by the impurity content. For comparison the conductivity of a typical good metal, Cu, is also shown. The arrows indicate the melting points.

impurity concentration low enough for the thermally excited electrons and holes to dominate the carrier concentration is said to be intrinsic, in contrast to the extrinsic, or impurity dominated regime.

Notice that the role of temperature and impurity content is quite different in metals from that in semiconductors. In semiconductors they control the conductivity through their effects on carrier concentration, and their role in producing sources of scattering is much less evident; in metals they dominate the conductivity as sources of scattering, and their effects on carrier concentrations can normally be neglected.

5.4   The dynamics of electrons and holes

The valence band of a semiconductor carries current only if some electrons, perhaps as few as 10^{13} cm^{-3}, have been thermally excited from it, for a weak electric field cannot affect a full band. In calculating the current carried by such a band it is easier to focus on the relatively small number of holes, rather than to work in terms of the large number ($\sim 10^{23}$ cm^{-3}) of remaining electrons. However, we must first establish the dynamical rules that are obeyed by holes.

Consider an otherwise full band with just one empty state at wave-vector k and also its complement, an otherwise empty band with this one state occupied (Fig. 5.6). The current carried by the first is to be regarded as that carried by the hole, j_h; when the two are added together a full band and zero current result, therefore

$$j_h = -j_e$$

The real space velocities of electron and hole are decided as usual by following the progress of a wave packet made up of states in the vicinity of k. The charge density of the packet associated with the single electron added to the charge scooped out by the single hole must result in the regular charge density associated with a full band; consequently the real space velocities of electron and hole are equal in magnitude and sign:

$$v_h = v_e$$

We can now use these equations to decide the charge q_h of a hole in terms of the charge q_e on an electron:

$$j_h = q_h v_h = q_h v_e = -j_e = -q_e v_e$$

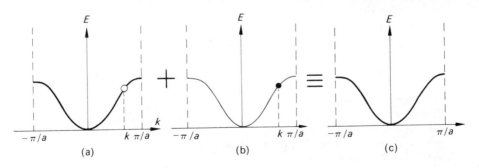

Figure 5.6   A band with a single electron state k empty (a), together with a band containing one electron only (b) is equivalent to a full band (c).

therefore

$$q_h = -q_e$$

A hole behaves as a positive charge.

The states of interest in a semiconductor are those near the bottom of the conduction band, and near the top (because holes decrease their energy by floating upward) of the valence band, and these are precisely the regions of band structure where Bragg reflections are important. We need to know the real space acceleration of an electron (or hole) under the influence of an electric field, for it is the balance between acceleration and scattering that determines the conductivity (§ 3.5), and if a diffracted wave is building up the free-electron equation of motion will not be adequate.

Consider first an electron; in an electric field $\&$ the equation of motion is (Equ. 3.9):

$$\hbar \dot{k} = e\&$$

and if that electron is near the top of a band the velocity (measured by the slope of the band) *decreases* with increasing energy instead of increasing as a free electron would (for example Fig. 5.4). It is useful to define a dynamical mass m^* by a Newtonian equation:

$$m^* \frac{dv_g}{dt} = e\& \tag{5.5}$$

into which we can insert the expression for the group velocity:

$$m^* \frac{d}{dt} \frac{1}{\hbar} \frac{\partial E}{\partial k} = e\& = \hbar \frac{dk}{dt}$$

or

$$m^* \frac{\partial}{\partial k} \frac{\partial E}{\partial k} = \hbar^2$$

so finally:

$$m^* (\partial^2 E / \partial k^2) = \hbar^2 \tag{5.6}$$

m^* is necessarily positive near the bottom of a band and negative at the top. There are no deep philosophical implications to a dynamical mass that is first positive and then negative; it is merely a parameterization that tells us how a wave packet moves when Bragg reflection is taken into account. A negative m^* means simply that the reflected wave is building up faster than the incident wave. Equ. 5.6 may be integrated twice (introducing two constants) to the form

$$E = \frac{\hbar^2}{2m^*}(k - k_0)^2 + E_0 \qquad (5.7)$$

which was that used in Section 5.2 for a band centred at k_0 and starting from energy E_0.

We can find the effective mass of a hole by following the real space acceleration of electron and hole wave packets. Since an electric field does not alter a full band the two packets must keep pace with each other and

$$\dot{v}_h = \dot{v}_e$$

The definition of m_h^* for a hole is obviously

$$m_h^* \dot{v}_h = q_h \mathcal{E}$$

and since the charge is opposite in sign to that of an electron, the effective mass must also have opposite sign to that of an electron in the complementary state:

$$m_h^* = -m_e^* = -\hbar^2/(\partial^2 E/\partial k^2)$$

The convenience of a description that uses both electrons and holes for a semiconductor is that only a relatively small number of carriers needs to be considered. Usually both electrons and holes have a positive mass because the former are at the bottom and the latter at the top of a band (there are exceptions because in three dimensions m^* is a tensor, not a scalar, and may have positive and negative components).

For our discussion it has not been necessary to define the wave-vector of the hole itself, the usual convention is to give it opposite sign to the wave-vector of the complementary electron. In that case if the total wave-vector of a system containing electrons and holes is required it is given by the algebraic sum of electron wave-vectors and hole wave-vectors.

5.5 Doped semiconductors

A semiconductor with a band gap of order 1 eV has very few carriers thermally excited at room temperature; a much more important effect on the carrier concentration is the presence of a small proportion of substitutional impurities that contribute a different number of electrons to the bands, for if there is any deficiency or excess of electrons the bands are no longer completely full or empty, and electrical conduction becomes possible. The important impurities are therefore those that will dissolve in the host and are either to the left or right of it in the periodic table, for example Al, P, Ga and As. We must now consider whether such an electron deficiency or excess is

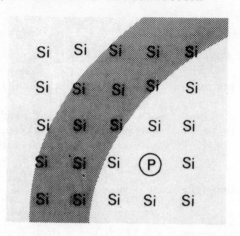

Figure 5.7   A single P impurity in a Si crystal has one extra positive ionic charge. There are bound states of the additional electron, analogous to those of a H atom (heavier shading).

produced in the *bands* or only in states localized on the impurity atom. Consider a P atom sitting in a Si host (Fig. 5.7); compared with the host the impurity (known as a donor impurity) has one more electron and the ion core carries one more positive charge; elsewhere the material appears electrically neutral, for within every atomic cell the fourfold ion core charge is balanced by the presence of four electrons. There is a Coulomb attraction between the extra electron and its parent ion, but one that is reduced by the dielectric constant κ of the host material; however a $1/r$ shape of potential is always capable of yielding bound states (that is not true of potentials that fall off more rapidly – for example, a sufficiently shallow square well has no bound states), so that the ground state of this system must resemble that of a H atom where the energy levels are given by:

$$E_n^H = -\frac{1}{n^2}\,\frac{e^4 m}{2(4\pi\kappa\hbar)^2} \equiv -\frac{1}{n^2}\,13\cdot6 \text{ eV}$$

but now the dielectric constant and also the effective mass alter the energy scale for the donor levels

$$E_n^D = \frac{m^*}{m}\,\frac{1}{\kappa^2}\,E_n^H$$

The dielectric constants of both Si and Ge are of order 10 and the effective masses of order $0\cdot1$ m, so that the energies of the donor levels are measured in hundredths of an electron volt. The spread of the wavefunction is also increased,

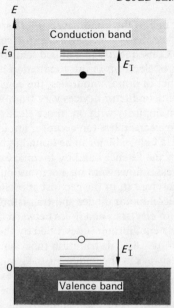

Figure 5.8   Impurity levels in a semiconductor. Note that for each donor atom an electron can occupy only *one* of the localized levels at any one time; likewise for a hole on an acceptor. Both electron and hole are shown in their ground states.

for that for H is characterized by the Bohr radius

$$a_0 = \hbar^2/(me^2)$$

which becomes

$$\kappa a_0 m/m^*$$

for the donor level, and is of order $20\,a_0$ for the ground ($n = 1$) state.

These bound states cannot, of course, contribute any conductivity but what happens to the donor electron when thermal energies are sufficient to ionize it, as they are at room temperature ($k_B T \simeq 0.025$ eV)? Ionization for a H atom means that the electron can escape from the proton to large distances and move freely through space. The analogue for a crystal is for the electron to join the conduction band, therefore *the ionization level must equal the energy of the bottom of the conduction band*, and the donor levels hang down into the gap (Fig. 5.8). For each donor impurity only one level can be occupied at any one time and the occupation probabilities are given by the Fermi–Dirac distribution function; the proximity of the conduction band with a great many available states means that although the likelihood of occupation of any one band state is always less than that for one of the donor levels, there are so many of them that even for $k_B T$ somewhat less than E_I the extra electron spends most of the time in the conduction band. Consequently in a doped semiconductor at room temperature the number of electrons in the conduction band is controlled by and nearly equal to the number of donor impurities; the concentrations that are used in the manufacture of semiconductor devices vary from 10^{13} to 10^{19} cm^{-3}.

If instead of adding an impurity with one more electron we add one, for example Ga, that has one electron less (an acceptor impurity), it can be thought of as contributing an extra hole, which can be bound in hydrogen-like levels, or ionized into the top of the valence band by thermal excitation (remember that the energy of a hole increases downward on a conventional diagram). The bound acceptor levels therefore sit in the gap just above the valence band.

The operation of semiconductor diodes and transistors depends on the equilibrium and transfer of electrons and holes between differently doped pieces of semiconductor. Particle equilibrium is described by the chemical potential μ which enters into the Fermi–Dirac distribution function. We have seen that in a pure semiconductor μ is always in the middle of the gap, but the addition of a single donor impurity at low temperatures (so that the extra electron is definitely in the lowest bound state) forces μ to lie above that level, since it is occupied with probability unity (Fig. 5.8). Similarly the addition of acceptor impurities forces μ down. At higher temperatures and concentrations the problem is more complicated, but it is generally true that a semiconductor containing donor impurities (n-type material) has μ near the top of the gap, and that with acceptor impurities (p-type material) has μ near the bottom of the gap. It is this difference in chemical potential that is the basis of the rectifying action of a p–n junction.

Notice that in compound semiconductors like InSb or CdTe a deviation from exact stoichiometry will have the same effect as doping with some impurity.

5.6   Compound semiconductors

In Section 2.4 we saw that the full bands and empty bands in the elemental semiconductors Si and Ge could be regarded as derived from the bonding and antibonding (sp^3) hybrid orbitals respectively. Since InSb and GaAs have related structures (cubic lattices with tetrahedral coordination of each group III atom by group V atoms and vice versa) it seems natural to use a related description of their band structures. However, if we consider the series of substances Ge → GaAs → ZnSe → CuBr the occupied s- and p-levels are increasingly associated with the less metallic component (As, Se, Br) and the empty s- and p-levels with the more metallic component (Ga, Zn, Cu). It is difficult to incorporate this qualitative view of electrochemical characteristics in simple Brillouin zone and energy band pictures, but some progress is now being made in this direction, especially by J. C. Phillips† and co-workers. In fact, compounds of the I–VII type have energy gaps too large for ordinary thermally activated semiconductivity to occur, but optical excitation by visible light photons ($\sim$2 eV) can be important; thus the *photoconductive* properties of AgBr and AgI are central to the photographic process. ZnS and CdS of the II–VI group of compounds are also photoconductive, but ZnTe and CdTe are semiconductors (with band gaps in the infrared), and for them it seems appropriate to associate the conduction band mainly with the 4s-levels of the Zn or Cd atoms and the valence band mainly with the 5p-levels of Te.

5.7   The optical properties of solids

When light quanta, photons, fall on a solid there is always the possibility of absorption. In this section we shall describe those processes which involve conduction or valence electrons; the interaction of photons with lattice vibrations is discussed in the book in this series by W. Cochran. If a photon is absorbed by a single electron, the difference between initial and final electron energies must equal the photon energy in order to ensure energy conservation:

$$\hbar\omega = E' - E$$

The spatial periodicity of the photon (wavevector q) imposes another constraint, for the electric field of the photon will have a spatial variation like $\exp(iq \cdot r)$, so that the matrix element for transitions between Bloch states of wave-vector k and k' looks like:

$$\int \psi_{k'}^* \exp(iq \cdot r)\psi_k d^3r$$

which can be written explicitly in terms of sums over plane waves:

$$\int \sum_G C'^*(G)\exp[-i(k' + G) \cdot r] \exp(iq \cdot r) \sum_G C(G)\exp[i(k + G) \cdot r] d^3r$$

An integral such as this taken over a macroscopically large volume vanishes unless there is a term with zero exponent, that is unless

$$k' = k + q + G$$

Not only is there energy conservation, but also wave-vector (or more loosely 'momentum') conservation. The velocities of electrons in solids are never more than about one-hundredth that of light, and so for a given energy the wave-vector of a photon is much less than that of conduction electron ($q = \omega/c$, whereas $k = 2E/\hbar v$ for free electrons). Since q is so small the initial and final electron states must lie almost vertically above one another in a (reduced) energy band

---

† See his book *Bonds and Bands in Semiconductors* 1974 Academic Press.

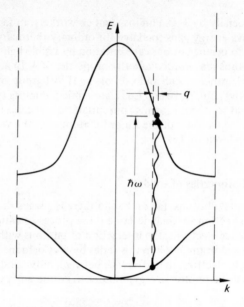

Figure 5.9   Absorption of a photon by a band electron.

picture (Fig. 5.9)†. Clearly there is a minimum photon energy and frequency for which such interband absorption can occur.

These single-particle transitions are usually dominant in the visible and near ultraviolet region of the spectrum, that is where the energies involved are up to about 10 eV. Further into the ultraviolet collective excitations of the electrons, known as plasmons, become important. However, in metals of complex electronic structure, the distinction between interband transitions and plasma excitations is not straightforward.

Optical spectroscopy of solids is becoming an experimental method of great importance because it probes the band structure well away from the Fermi level, unlike measurement of the Fermi surface itself. In insulators and semiconductors the optical absorption provides a direct measure of the energy gap between the top of the valence band and the bottom of the conduction band for, since the bands are either completely full or completely empty, no intraband transitions

† It might appear from Fig. 5.9 that low frequency photons, say in the microwave or radiofrequency range, cannot be absorbed by a metal for there are no final states within reach; however, the bands are not completely sharp because scattering of electrons blurs the wave-vectors by an amount of order $1/l$ where l is the electron mean free path, and transitions can take place. An alternative description of these low frequency processes is to think of them in terms of a frequency-dependent conductivity.

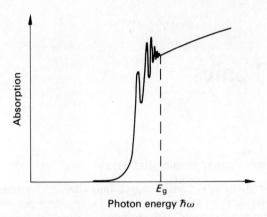

Figure 5.10   Absorption coefficient of a semiconductor as a function of photon energy (schematic).

can occur. However, experimentally the onset of absorption is not completely devoid of structure (Fig. 5.10), instead there appear to be a number of energy levels looking rather like a series of spectral transitions in a simple atom. The reason is that the gap energy is that energy required to create a hole at the top of the valence band and an electron at the bottom of the conduction band, which can then travel independently throughout the crystal; however, some saving in energy may be made by keeping the two close together and taking advantage of the Coulomb attraction between them. The states of this electron–hole pair will be hydrogen-like and resemble those of an impurity in a semiconductor (Fig. 5.7); the energy scale and spatial extent of the wavefunction will depend on the effective masses of electron and hole and also the dielectric constant. Such an excitation is known as a Mott exciton.

Another type of exciton exists that can be created by optical excitation in insulating compounds with both hole and electron on the same ion, atom or molecule, but the excitation can jump to a neighbour and so travel through the crystal; these are Frenkel excitons.

There exist also in insulating compounds a whole set of interesting and optically active electronic phenomena that are associated with crystal defects, of which the best known is an electron trapped at a negative ion vacancy – the F centre. A full account of these features, which have considerable practical importance, is given in the companion text in this series *Defects in Crystalline Solids* by B. Henderson.

6

Special Topics

6.1   Introduction: Beyond simple Bloch states

In most of the discussion of the two preceding chapters we have been considering the distribution and occupation of independent one-electron energy levels in a periodic crystal lattice. This approach to the electronic structures and properties of solids is a very powerful one and provides the most useful entry point to many problems. There are, however, a number of striking physical phenomena for which this approach is inadequate, and for which further considerations must be injected into our physical picture of electrons in solids, and we shall deal with some of them in this chapter. Superconductivity provides an obvious example, but there is, however, a much more central problem, connected with the distinction we have made between metals and insulators. In terms of Bloch states in Brillouin zones perfect insulators are rather a special case. They require (see p. 108) an even integral number of electrons per primitive unit cell, and band gaps large enough to ensure that no overspill takes place from the highest filled zone into the next zone. Such an insulator carries no current because there are no empty states within reach of the occupied states, and for the same reason the material is transparent to radiation whose quantum energy is less than that of the band gap.

We have seen that there are materials well described by this picture. However, as we saw in Section 5.5 the addition of a few electrons to an insulator by adding, for example, a few atoms of As to Si, does not *at 0 K* immediately yield the conductivity that would follow if the only allowed states for these extra electrons were Bloch states. Much more strikingly, there are insulators that certainly do not become metallic conductors when melted, despite the fact that Brillouin zones are a consequence of crystalline regularity and should disappear at the melting point. With the Bloch state—Brillouin zone model all liquids should be metallic conductors and none should be transparent; similar arguments should apply also to amorphous materials such as ordinary glass. This conflict with everyday experience shows clearly that something is missing from the theory.

The question can be put the other way around, and we can ask whether all materials with one electron per unit cell are metals. For example, if we took metallic crystalline Na and could somehow steadily increase the lattice spacing the simple theory would require it to stay metallic, but common sense suggests that an array of Na atoms at a macroscopic spacing could not conduct electricity.

What we have omitted from the model (or at least failed to include explicitly) are the interactions and correlations between electrons. We have required that on a large scale a solid should be electrically neutral, the number of conduction electrons being balanced by the positive ion core charges. On a small scale this allows fluctuations in local charge densities as if the electrons were independent. We pointed out for the hydrogen molecule problem in Section 2.2 that the simplest molecular orbital approach is incorrect in making configurations in which both electrons are on one atom too probable (as compared with those where one electron is on each atom). In fact the second electron does not simply see the time average of the charge density of the first electron, but moves in a correlated fashion arranging to spend most of its time in other parts of the molecule than those occupied by the first. We must now try to take such correlations into account in a solid. We shall first see how electrons behave in disordered materials and then look briefly at the question of the metal–insulator transition.

6.2   Disordered materials and the metal–insulator transition

We have already looked at two kinds of disorder, metallic alloys (§ 4.7) and doped semiconductors (§ 5.5), and the reader will have noticed that the two were treated differently. When a Zn atom is put into a Cu crystal the doubly charged Zn ion does attract an additional charge density into its vicinity, but this charge density is made up of a lot of small contributions from a wide range of conduction states; no single electron is bound to the Zn impurity. On the other hand for P impurities in Si such bound states do exist. The difference comes about from the form of the potential in the two cases (Fig. 6.1); in a metal the mobile conduction electrons screen out a Coulomb $1/r$ potential very rapidly, and at large distances the screened potential drops off exponentially. A semiconductor is quite polarizable (the dielectric constants of Si and Ge are both of order 10), but the electrons are not mobile; the screened potential therefore has the form $1/(\kappa r)$, which falls off sufficiently slowly for a bound state to form.

If we take a semiconductor with a large concentration N of impurities there is an interesting situation at low temperatures. First of all each impurity might be in its own ground state, but if these ground state wavefunctions overlap each other there is the possibility that electrons from neighbouring impurities can stray on to a given site and start to screen its potential more strongly, and so release its electron. This effect is a cooperative one, for the more electrons that are released and become mobile, the stronger the screening of the impurity potential and the more difficult it is for any electron to remain bound.

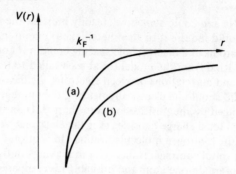

Figure 6.1   The potential for electrons around a positively charged impurity: (a) in a metal; (b) in a semiconductor. The fast exponential decay in a metal has characteristic length of about k_F^{-1}.

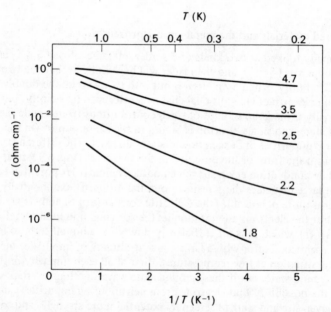

Figure 6.2   The very low temperature conductivity of heavily doped Ge. Notice the enormous change in conductivity for a small change in impurity concentration, which is marked on each curve in units of 10^{17} cm^{-3}. After F. R. Allen and C. J. Adkins, *Phil. Mag.* **26**, 1027 (1972). Compare with Fig. 5.5.

The critical condition for this to occur can be estimated by asking what would happen to the impurity potential if all the donor electrons were mobile. The form of the potential is then (see references for this chapter)

$$V(r) = -\frac{1}{\kappa r} \exp\left(-\frac{r}{l}\right)$$

with l, the characteristic screening length, given by

$$l^2 = \frac{\hbar^2 \kappa}{4m^*e^2} \left(\frac{\pi}{3N}\right)^{1/3} \simeq \frac{\hbar^2}{4m^*e^2} \kappa d$$

where d is the average distance between impurities. For distances much smaller than l, the $1/r$ potential is retained, so that if the ground state wavefunction, calculated with a $1/r$ potential, had a radius considerably less than l, screening would hardly affect it, and the bound state situation would be stable. If the radius of the wavefunction were greater than l, the set of bound states would be unstable against this 'melting' of the electrons.

The radius of a hydrogenic ground state for a donor in a semiconductor is

$$a = \frac{m}{m^*} \kappa a_0$$

where a_0 $(= \hbar^2/me^2 = 0.53$ Å$)$ is the H atom Bohr radius, so that the condition becomes:

Insulator: $d \geqslant 4\left(\frac{m}{m^*}\right) \kappa a_0$;

Metal: $d \leqslant 4\left(\frac{m}{m^*}\right) \kappa a_0$

The numerical values for Ge indicate a critical spacing of about 100 Å or a critical concentration of about 10^{18} cm^{-3}. At low concentrations (and low temperatures) all donor electrons are in bound states, at high concentrations these electrons can move from site to site at any temperature and behave rather like a free-electron gas. The measured resistivity of heavily doped Ge does indeed show this change from an insulator to a conductor (Fig. 6.2) as the impurity concentration is increased, and the rapidity of the change shows that it must be a collective phenomenon. In the conducting regime the electrons are said to form an impurity band.

Although we have obtained this result for a random distribution of centres the idea applies just as well to a crystal; if we could expand sodium metal (or anything else with hydrogenic atoms), such an approach would yield an insulator for interatomic spacings greater than about 2 or 3 Å (the dielectric constant and m^*/m are now unity because there is no background medium). The usual nearest-neighbour distances in metals are 2·5 to 3 Å, so the precise numerical value of this criterion should not be taken too seriously, and other theoretical approaches do give somewhat different numbers. However, the resistivity behaviour under pressure of some metals, for example Cs, indicates that they are not very far from the metal to non-metal transition, and these are metals of particularly low density — the interatomic spacing of 5·23 Å in Cs is the largest of all metals.

A large number of transition metal oxides, sulphides and selenides (known collectively as chalcogenides) undergo a metal–insulator transition when the pressure or temperature or both are varied. The physics of the transition in these materials is complicated by the additional possibility of changes in magnetic order at the transition, but certainly involves also the ideas we have just discussed. The whole subject has been very clearly treated in the book by Mott referred to at the end of this book.

6.3   Liquids and glasses

The keystone of the nearly-free-electron approach to simple metals is that the interaction with the positive ions (or more precisely, the effective interaction and the pseudopotential) are small. For that reason not only are the band gaps and distortion of the Fermi surface small, but also the result of displacing the ions by thermal fluctuations is also small, so that only a small resistivity is contributed by the phonons. At room temperature the mean free path of an electron in simple metals is usually of order 10^2 lattice spacings, even though the typical amplitude of thermal vibration of the ions is then about a tenth of a lattice spacing.

On melting all long-range order is lost, and with it the possibility of diffraction, but the density and typical interatomic distances change but little. Prior to melting only a small proportion of the conduction electrons is affected by the long-range order, only those that are close to the Bragg condition for diffraction (if the proportion were not small the metal would not be simple in our sense). Consequently, melting should hardly affect most of the electrons, and the density of states and the electronic mean free path should not greatly change. That these ideas have some validity is shown most directly by a comparison of the electrical resistivity in solid and liquid phases at the melting point (Table 6.1); the resistivities of the liquid simple metals are not that much greater than those of the solids, so that the electronic mean free path must still be at least tens of interatomic spacings.

On the other hand, the resistivity of Si and Ge drops on melting (Fig. 5.5), as also does that of the semi-metal Sb, and in the liquid state these materials have a resistivity that increases with temperature, unlike solid semiconductors. In these elements it would appear that the band gaps have indeed been destroyed by melting and that the liquids are true metals.

However, if the same kind of description is applied to liquid Ba, and the free-electron formula is used for the conductivity, the measured value corresponds to a mean free path of about 5 Å, which is of the same order as a lattice spacing (and also the de Broglie wavelength of the electrons). The inference is that the scattering is so strong that it is meaningless to start from a description in terms of free electrons. High resistivities, of order 10^3 $\mu\Omega$ cm, in liquid or solid always indicate a short mean free path, and must therefore signal the breakdown of a free-electron approach and departures from simple metallic behaviour. Indeed, these resistivities represent an upper bound (the Mott limit) for metallic behaviour, defined by a resistivity that becomes independent of temperature at low temperatures. On the non-metallic side of the Mott limit interactions and correlations between electrons must be made explicit, and can no longer be incorporated into a weak effective pseudopotential.

There are plenty of materials with liquid or amorphous phase resistivities that are well above the Mott limit, and amongst them glasses have great technological importance. (Glasses are liquids of very high viscosity; when heated they soften gradually but do not have a melting point; some glasses can be persuaded to crystallize and become true solids at low temperatures.) They should therefore be approached from a direction diametrically opposite to the free-electron standpoint, and electron correlations should be built in from the start. (Another simple example of strong correlations where insulating properties are preserved in the liquid state is provided by the low density liquid inert gases.)

Suppose then we create our glass by starting with atomic wavefunctions on well separated atoms and then bring the atoms together. The wavefunctions start to overlap and interact, and, if we are dealing (as we are with most glasses) with a covalent material, bonds are formed between the atoms that are well defined in direction and length and contain high electron density. A covalent bond between two atoms involves sharing of a pair of electrons, one each of spin-up and spin-down between them. A covalent bond in a molecule is the analogue of a band for a metallic solid, when fully occupied both are inert and difficult to disturb; an unsaturated bond (in chemical language) is reactive, just as the distribution of occupied states in an unfilled band can easily be changed.

Table 6.1   Electrical resistivities of solid and liquid at the melting point.

Element	Na	K	Cu	Mg	Ba	Al	Pb	Sb
Melting point temperature (K)	371	339	1356	924	983	933	600	904
$\rho_s(\mu\Omega$ cm)	6·6	8·3	10	15	80	11	49	183
ρ_l ($\mu\Omega$ cm)	9·6	13	21	28	306	24	85	109

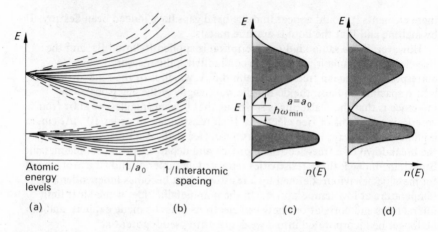

Figure 6.3   The electron energy levels in an element (a) that broaden into bands in an amorphous solid (b); the solid lines represent states that are delocalized and can carry current, they resemble Bloch states; the broken lines correspond to localized non-current-carrying states (cf. Fig. 5.8) that lie within the Bloch state energy gap E_g; (c) the resultant density of states; Bloch-like states are heavily shaded, localized states lightly shaded; (d) situation in which there are localized states at all energies within the original gap.

The chemist's bonding picture is independent of crystallographic order, for it is the number of bonds to each atom, four to Si and two to O and so on, that is important, and it will usually be possible to satisfy the bonds with either a regular or an irregular arrangement of atoms. If all bonds are saturated then all available electrons are locked into position and the material is an insulator.

The two kinds of picture, bands and bonds, are representative of the extreme situations, a completely free-electron metal on the one hand, (say Na), and a strongly covalently bonded material, say SiC, on the other. In between it may be possible, and sometimes necessary, to utilize both viewpoints. So for elemental Si, the crystal structure involves the fourfold tetrahedral bonding arrangement, and the measured electron density certainly shows such bonds, but on the other hand a band picture is most useful for describing excitations and the effects of impurities. We can therefore ask a band type of question of a bonding situation, and look at the distribution of allowed states with energy for a glass (or a liquid or an amorphous material). We start off with well defined atomic levels (Fig. 6.3a) that broaden as they are brought together (cf. Fig. 4.23). From a band point of view an energy gap must exist for an ordered tetrahedral structure because of diffraction; in a disordered material there cannot be any sharp cut-off and some states are found spilling into the gap (Fig. 6.3b). However, these states are not running wave states but are instead localized and incapable of carrying current; this situation is the natural development of what happens in a doped crystalline semiconductor: the addition of an impurity

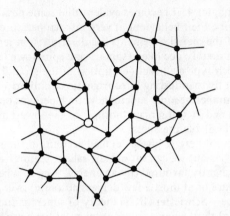

Figure 6.4   Bonding of amorphous 4-coordinated solid, with one 5-valent impurity (0). The extra bond is easily taken up because of the irregularities in the network.

introduces states into the gap, but these states are localized near the impurity and are not current carrying.

Although the localized states of the band tails carry no current they certainly do provide states from which optical transitions may be made, so that the absorption edge will correspond to a photon energy that bridges the tails (Fig. 6.3c). That there is a reasonably well defined absorption edge in amorphous materials indicates that the density of states in the tails drops very fast near the centre of the gap, and in fact there is an exponential decay (Mott and Davis, see **Further Reading**). In some materials the localized states do manage to fill the gap (Fig. 6.3d) so that there is no optical gap at all, but on the other hand the material is still an electrical insulator.

A special feature of amorphous semiconductors and insulators is that they are insensitive to the addition of impurities, quite unlike crystalline semiconductors. The reason for this is that in an amorphous material it is the bonds that matter, and the network can easily take in an atom with one extra or one less bond (Fig. 6.4). The additional electron is well localized within this bond and, because of the absence of translational symmetry, does not spread out over tens of lattice spacings in the way that a donor state in a semiconductor crystal does (§ 5.5). Consequently the energies required to move an electron around in an impure amorphous semiconductor are not particularly small, and the electrical conductivity is not greatly affected.

6.4   A note on superconductivity

Even for many of those elements that are, at room temperature, metallic conductors, and appropriately described therefore by the Bloch state-Brillouin zone language of Chapter 4, it becomes evident that some new concepts must be introduced to describe their behaviour at very low temperatures. It turns out that for very few of the metallic elements is a description in terms of a Fermi surface in a band of doubly occupied (spin-up and spin-down) single-particle energy levels of Bloch type a correct description of the ground state. A few of them depart from it by manifesting ferromagnetic or antiferromagnetic spin ordering: a larger number become superconducting; perhaps only the alkalis, and Cu, Ag, Au, Rh, Pd and Pt will remain 'normal' to the lowest attainable temperatures at all available pressures.

We shall not attempt here to give a detailed account of the superconducting state and its properties but we can set out the salient features of the new ground state which is very slightly favoured (by an energy $\sim k_B T_c$ where T_c is a transition temperature of at most a few degrees absolute) over the normal state. The Bardeen–Cooper–Schrieffer (BCS) theory of superconductivity is built around two facts: (a) that, when account is taken of the coupling between electrons and the zero-point motion of the ions, an attraction between two electrons by the emission and absorption of virtual phonons can exceed their screened Coulomb repulsion; and (b) that in the presence of such an attraction the normal state is unstable with respect to the condensation of electrons at the Fermi surface into pair states (Cooper pairs) the energy gain being greatest when when the components of the pair are $k\uparrow$ and $-k\downarrow$. In a normal metal states of different k are occupied with independent probabilities given by the Fermi–Dirac distribution function, in a superconductor the occupancies are highly correlated, and the states are occupied pair-wise. The attraction can be envisaged as a polarization of the lattice of positive ions by one electron of the pair such that the energy of the other is lowered by occupying that region, rather as one ball-bearing on a rubber sheet will produce a depression towards which a second will move. The 'condensation' of the Fermi surface into pair states takes place in a cooperative fashion and at 0 K no conventional single-particle states exist in an energy shell of thickness about $3 \cdot 5\, k_B T_c$ around the Fermi surface. There is thus an energy gap for single-particle excitations that is slightly reminiscent of the situation in a semiconductor. One should note, however, that the gap is tied to the Fermi surface (rather than to Brillouin zone boundaries) and also that the single-particle excitations ('quasi-particles') in a superconductor at finite temperature are not simple Bloch states, because they still feel the effects of coupling to other electrons.

The prevalence of superconductivity among the metallic elements has already been indicated and clearly no very special features are required in the normal state electronic structure for it to manifest itself. The BCS expression for the transition temperature,

$$k_B T_c = \hbar\omega_D \exp(-1/n(E)_F V)$$

involves, as well as the characteristic lattice frequency ω_D and the net attractive interaction V, the density of electron states $n(E)_F$ at the Fermi energy, which determines the number of electrons forming pairs. However as with many other properties of metals where this density of states plays a role, it cannot carry the burden of representing *all* the relevant features of the electronic structure.

Further Reading

General
Introduction to Solid State Physics. C. Kittel, John Wiley and Sons Inc., New
    York and London, 4th Edition 1972.

Chapter 1
Atomic Structure and Atomic Spectra. G. Herzberg, Dover Publications.
Elementary Wave Mechanics, N. F. Mott, Wykeham Publications, London 1973.

Chapter 2
Valence, C. A. Coulson, O.U.P., Oxford 1952.
Quantum Theory of Matter, J. C. Slater, McGraw-Hill, New York 1953.
Structural Inorganic Chemistry, A. F. Wells, Clarendon Press, Oxford 1962.

Chapter 3
The Theory of the Properties of Metals and Alloys, N. F. Mott and H. Jones,
    O.U.P., Oxford 1936 (Dover reprint).
The Wave Mechanics of Electrons in Metals, S. Raimes, North-Holland,
    Amsterdam 1961.

Chapter 4
Electrons in Metals, J. M. Ziman, Taylor and Francis, London 1963.
Quantum Theory of Solids, R. F. Peierls, Clarendon Press, Oxford 1955.

Chapter 5
Atomic Theory for Students of Metallurgy, W. Hume-Rothery and B. R. Coles,
    Inst. of Metals, London 1969.
An Introduction to Solid State Physics and its Application, R. J. Elliott and
    A. F. Gibson, Macmillan, London 1974.

Chapter 6
Electronic Processes in Non-Crystalline Materials, N. F. Mott and E. Davis,
    Clarendon Press, Oxford, 1971.
Metal-Insulator Transitions, N. F. Mott, Wykeham Publications, London 1974.
Superconductivity, E. A. Lynton, Methuen, London 1971.

Index